C0-ATI-104

MGEN
NE94

WIRELESS NETWORKING

A How-To-Do-It Manual for Librarians

LOUISE E. ALCORN
MARYELLEN MOTT ALLEN

Université d'Ottawa
BIBLIOTHÈQUES
University of Ottawa
LIBRARIES

HOW-TO-DO-IT MANUALS
FOR LIBRARIANS

NUMBER 131

NEAL-SCHUMAN PUBLISHERS, INC.
New York, London

b29406808

Published by Neal-Schuman Publishers, Inc.
100 William St., Suite 2004
New York, NY 10038

Copyright © 2006 Neal-Schuman Publishers, Inc.

All rights reserved. Reproduction of this book, in whole or in part, without
written permission of the publisher, is prohibited.

Printed and bound in the United States of America.

The paper used in this publication meets the minimum requirements of American National Standard for Information Sciences – Permanence of Paper for
Printed Library Materials, ANSI Z39.48–1992.

Library of Congress Cataloging-in-Publication Data

Alcorn, Louise E., 1970-
    Wireless networking: a how-to-do-it manual for librarians/Louise E. Alcorn,
Maryellen Mott Allen.
        p. cm. – (How-to-do-it manual for librarians; no. 131)
    Includes bibliographical references and index.
    ISBN 1–55570–478–6 (alk. paper)
 1. Telecommunication in libraries. 2. Wireless communication system.
3. Library information networks. 4. Public libraries—Information technology.
5. Academic libraries—Information technology. L. Allen, Maryellen Mott,
1970- II. Title III. How-to-do-it manuals for libraries; no. 131.
Z680.5.A43 2006
025'.00285—dc22   2004053760

Z
680.5
A43
2006

For my mom,
who would love me if I never wrote a word,
but who is nonetheless delighted that I got this done.
And to the WF. You keep me sane.
Louise

For my friends and family.
Maryellen

CONTENTS

LIST OF FIGURES

PREFACE

Wireless networks in libraries, like other technological improvements of the past quarter century, are no longer innovations, but expected services. Both of us have been involved in the successful transition from wired to wireless, Maryellen at a large academic institution and Louise at a busy mid-size suburban public library. During the migration, we searched for books that addressed both professional interests and practical concerns about wireless networking. There is a plethora of books available on the general topic, ranging from "idiot's guides" to technical treatises, yet all of them left us unsatisfied. What was missing was an easy-to-understand guide that addressed library-specific needs. We hope you will find *Wireless Networking: A How-To-Do-It Manual for Librarians* to be just that.

Libraries come in countless shapes and sizes, and each one faces its own unique set of circumstances. In addition to individual facilities, each institution also serves a distinct user group or groups. Because all libraries face their own budget constraints, any new technologies must be approached with these in mind. While most libraries do not typically charge for their services, their patrons still demand that they keep pace with technological innovation. Although a "one-size-fits-all" guide would be impossible, there are still numerous points of correspondence between institutions. This practical guide takes into account all of these circumstances and tries to provide appropriate, timely guidance.

The idea that a librarian may be called upon to implement a wireless network is no idle fantasy. Many small public libraries and school libraries have only one person in charge of all aspects of the library, including technology. Even if your library, like many modern organizations, has staff whose sole responsibility is managing technology, it is important that all members of the team have a basic understanding of how their institution's systems work. For this reason, we've tried to explain technologically intense concepts in reasonable layperson's terms, building upon the essential concepts of local area networking. We focus on expanding your network capabilities to include wireless accessibility. "The Wireless Sourcebook" (pages 131–193) also includes some excellent resources on basic networking concepts for those in need of further information.

ORGANIZATION

Wireless Networking: A How-To-Do-It Manual for Librarians begins by explaining the fundamental ideas behind wireless technology, moves through reasons to "go wireless," considers planning issues, and compares hardware choices. Complex topics are also examined, including security, maintenance, and troubleshooting policies. Throughout the guide, real-life librarians' experiences are interspersed with our explanations, helping to anchor the concepts.

Chapter 1, "Wireless Networking Basics," provides a brief history of the technology, an explanation of how it works, and typical operations. The various types of networks—personal, local area, metropolitan, and wide area—are also explored.

The next chapter, "Why Go Wireless?" makes the case for taking on this project. We explain how wireless can aid in point-of-need services, improve technology in large spaces, and eliminate the restrictions of older or historic buildings. We also delve into the future of this technology in libraries.

Chapter 3, "Planning for Wireless Networks," takes on one of the most important topics. Nothing derails a technology project faster than poor or inadequate planning. We cannot stress enough how vital it is to know what you are doing, who is responsible for different aspects of the project, how long the process will take, where technology will be placed, and how it will be implemented. Special topics, including policy writing, communication, and marketing, are also discussed.

Now that you have made your plan, it is time to start narrowing in on the specifics. Chapter 4, "Hardware Equipment and Installation," will help you make appropriate choices by covering factors for selection and options. This chapter includes information on choosing equipment and guides to pricing. Maintaining your network is an easy task if you are aware of the main issues you face.

Chapter 5, "Securing, Maintaining, and Troubleshooting Wireless Networks," addresses many of the most common security and maintenance matters and suggests proactive ways to help your library avoid troublesome concerns. We outline security concerns and discuss options for protecting your network, and offer suggestions for troubleshooting your equipment and advice for creating FAQs and help documents for staff and users.

Technology projects, especially those that bring us to new territories, can be frustrating and often lonely undertakings. Since you may be the first library in your area to take this project on, you may not have anyone else's experience from which to learn firsthand. Our "User Experiences" sections in shaded boxes throughout the text, which share advice and suggestions from all types of libraries, will ameliorate feelings of isolation, showing you that you are not alone in your trials and triumphs.

Finally, "The Wireless Sourcebook" provides a variety of helpful tools. The "Glossary of Terms" clarifies new terms used in this book while the "Read More about It" section offers sources for further information. "Sample Policies and FAQs" gives you documents to model, and "Equipment Manufacturers" offers a list of vendors to consider. In the last source, we offer you excerpts from an actual wireless site survey, a vital tool for any wireless implementation, courtesy of the Maricopa County Library District in Arizona.

We hope that *Wireless Networking: A How-To-Do-It Manual for Librarians* proves a uniquely useful resource for those working in our profession. As you and your library move to better serve your staff and users, wireless networks will prove a valuable addition to your current technology plan.

ACKNOWLEDGMENTS

MARYELLEN

I would like to thank the faculty and administration of the University of South Florida Tampa Library for their support.

LOUISE

I would like to thank the members of the Web4Lib, LibRef, Publib, Iowalib, and Libwireless e-mail lists, for invaluable contributions of their own experiences with wireless—good, bad, and otherwise. The wealth of knowledge and easy generosity of the library community continue to astonish and delight me. Many, many grateful thanks!

One particular bit of thanks to Graceanne DeCandido—for her clear, well-written articles, several good leads, and specifically for a delightful Sunday afternoon e-mail exchange during which she cooked and I stewed (over this book). And for reminding me to "make sure to tell the bit about Hedy Lamarr!" Grazie.

Another huge, personal thank-you to Rachel Singer Gordon, for a wonderfully productive afternoon in a Starbucks, and for her wry wit and clear head about writing projects.

Many, many thanks to Paula Wilson and Vicki Terbovich of Maricopa County for helping me down the home stretch. Ladies, you're truly fabulous! Vicki, I owe you a drink. Or twelve.

As always I thank my family, especially my brother, Justin, who put me on the track of some security issues.

Finally, I once again thank my colleagues at the West Des Moines Public Library in Iowa—for time, resources, and dependable Internet access. And for not laughing at me when they found me face down, asleep at my desk. Twice. Cheers.

1 WIRELESS NETWORKING BASICS

OVERVIEW
• **A Brief History of Wireless Technology** • **The Electromagnetic Spectrum** • **What is Wireless Networking?** • **Wireless Generations** • **Typical Operation of Wireless Networks** • **Types of Wireless Networks**

In this chapter, we hope to give you a simple primer on wireless technology—its history, its basic functions, and where it's going. In reality, frequency modulation, transmission, and spreading technologies are much more complex than the brief overview given here. However, we feel the information contained here is a good, basic grounding to get you started with wireless implementation. In the same vein, this is not an introduction to networking in general, though we do discuss the basic workings of wired and wireless networks. If you need a more technical explanation, or simply want to learn more, please consult one of the excellent resources cited in "The Wireless Sourcebook" at the end of this book or refer to Source A, which provides a complete glossary of the terms used in this book.

A BRIEF HISTORY OF WIRELESS TECHNOLOGY

While the history of the development of wireless is not essential to the installation of a network, a little background may prove interesting to you. If you are a devotee of those British period mystery dramas on PBS, you've likely heard radio referred to as "wireless." In point of fact, when we speak of "wireless," we're talking about radio. When we discuss wireless as a means of *data* transfer, the forms of transmission will vary, but the basic concept is effectively the same as with radio.

Like most aspects of the history of science, it can be difficult to point to a specific starting point at which the development of wireless began. Lewis Coe, in *Wireless Radio: A Brief History,* begins the story of wireless with the Scottish inventor and physicist James Clerk Maxwell (1831–1879). Maxwell was a serious student of electromagnetism who developed equations in electricity and magnetism (1873) that described the behavior and relationship between the two fields. His formulas indicated that there were electromagnetic waves in the same way there were light waves.

Maxwell did not carry out any practical experiments, but some thirteen years later Heinrich Hertz (1857–1894) was able to conduct laboratory experiments with electromagnetism. Coe notes that "the idea of communicating through space was so novel at the time that most serious scientists did not dwell on it or even consider it." (Coe 1996, 4) Guglielmo Marconi (1874–1937) followed up on Hertz's experiments. Working in the attic of his family's home on their estate in Bologna, Marconi worked to send transmissions across longer and longer distances. By 1895, at age twenty-one, Marconi had succeeded in sending transmissions across a mile and a half. He also proved that it was possible to send transmissions in spite of obstacles by setting up a transmitter near his house and a receiver three kilometers away, behind a hill.

Marconi approached the Italian government with his invention, but there was no interest. Instead, the British provided him with the support he needed to continue developing practical applications of wireless. In 1897, Marconi finally received a patent (# 12,039) for "wireless" and established the Wireless Telegraph and Signal Company. He continued to work to extend the distance of transmissions, and in 1898, Marconi successfully transmitted signals across the English Channel. He received a Nobel Prize in 1909 for his work in radio frequencies and wireless technology. By 1927, a worldwide short-wave radio network was in use.

Wireless was originally most important in military communications. The Royal Navy used wireless to communicate with ships at sea, and the British quickly became leaders in the use of radio technology in armed forces communications. Indeed, the use of radio revolutionized military communications, allowing for unprecedented speed and accuracy in the transmission of strategic information.

THE ELECTROMAGNETIC SPECTRUM

There is no real need for the general reader to have an in-depth understanding of the electromagnetic spectrum in order to work

Electromagnetic spectrum
The electromagnetic spectrum describes the range of all forms of radiation from sound to radio waves, through visible light, to harmful x-rays and gamma rays. Different sections of the spectrum are called bands, containing a range of frequencies within the spectrum that can be classified, such as visible light or infrared radiation.

Bands (radio bands)
A band is a small section of the electromagnetic spectrum of radio communication frequencies, in which channels are usually used or set aside for the same purpose.

Wave
The basic component of the electromagnetic spectrum, waves (such as radio waves) travel and transfer energy from one point to another. Waves are characterized by crests and troughs, either perpendicular or parallel to wave motion.

Wavelength
The distance in the line of advance of a wave from any one point on a wave to a corresponding point on the next wave.

Frequency
The measurement of the number of times that a repeated event occurs in a fixed period of time, measured in Hertz (Hz). With radio, the frequency is the number of cycles of the repetitive waveform per second.

Hertz
A unit of measure of electromagnetic frequency named after Heinrich Hertz. One hertz corresponds to one cycle per second (or one wave per second).

with wireless technologies. However, a basic overview of the spectrum can be quite helpful in understanding these technologies. The **electromagnetic spectrum** describes the range of all forms of radiation from sound to radio waves, through visible light, to harmful x-rays and gamma rays. Different sections of the spectrum are called **bands,** containing a range of frequencies within the spectrum that can be classified, such as visible light or infrared radiation. In reality, there exists no actual boundary between the bands and there is much overlap between frequencies, but the divisions allow for greater ease in classifying and utilizing the energies.

Almost everything gives off radiation in some manner. Even the human body emits radiation in the form of heat (part of the infrared spectrum). It is important to understand that all parts of the spectrum are similar in that they are all electromagnetic energy. The differences lie within the varying frequencies of the energy.

The **wave** is the basic component of the electromagnetic spectrum and is described by its **wavelength** and **frequency**. The *wavelength* is the distance between the crests of a wave and is measured in meters. The *frequency* refers to how many waves travel through a medium (such as air) over a given amount of time and is measured in cycles-per-second, or **Hertz** (Hz). The wavelength and frequency of a wave are linked together. The longer the wavelength, the lower the frequency while the shorter the wavelength, the higher the frequency. This makes sense in that shorter wavelengths will hit an antenna more often because the crests of the waves are closer together.

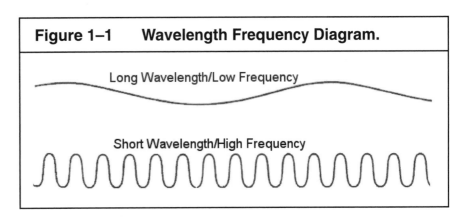

Figure 1–1   Wavelength Frequency Diagram.

Long Wavelength/Low Frequency

Short Wavelength/High Frequency

Although the spectrum is theoretically infinite, we will discuss the most commonly used range of frequencies. At the low end of the spectrum, we find radio and microwaves. These waves have low frequencies, low energies, and long wavelengths. Measured in meters, the size of a radio wave (from crest to crest) at a frequency of 750 kilohertz would be about four hundred meters. At the other end of the spectrum, x-rays and gamma rays are exceedingly small and are measured in trillionths of a meter.

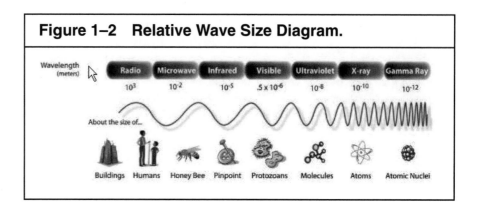

Figure 1–2   Relative Wave Size Diagram.

The portion of the electromagnetic spectrum in which wireless networks reside is toward the low end, within the bands that are defined by radio. See the chart below for a more detailed explanation.

Figure 1–3   Electromagnetic Frequencies and Applications.

Frequency (Hz)	Type	Applications
10^{15} +	**Gamma Rays, X-rays**	Scientific/Medical
5.0 (10^{14}) – 7.5 (10^{15})	**Ultraviolet**	Scientific/Medical
4.3 (10^{14}) – 7.6 (10^{14})	**Visible Light**	
3.0 (10^{11}) – 4.3 (10^{14})	**Infrared**	Photography/Physical Therapy/IR Remote Devices
10^{9} – 3 (10^{11})	**Microwaves**	Radar/Navigation/Microwave Ovens/ LoS Networking
10^{9} – 4.3 (10^{9})	**Radio**	Local Multipoint Distribution Service
.25 (10^{9}) – .27 (10^{9}) .34 (10^{9}) – .37 (10^{9})	**Radio**	Multichannel Multipoint Distribution Service
.515 (10^{9}) – .535 (10^{9}) .5725 (10^{9}) – .5825 (10^{9})	**Radio (Unlicensed)**	National Information Infrastructure (802.11a)
.24 (10^{9})	**Radio (Unlicensed)**	802.11b/HomeRF/Bluetooth/Consumer Electronics (Cordless telephones)
.185 (10^{9}) – .181 (10^{9}) .193 (10^{9}) – .199 (10^{9})	**Radio (Licensed)**	Cellular Personal Communications System (PCS)
.191 (10^{9}) – .193 (10^{9})	**Radio (Unlicensed)**	Cellular Personal Communications System (PCS)
.12 (10^{9}) – .13 (10^{9})	**Radio**	Mobile Satellite and Global Positioning System (GPS)
93.5 (10^{8}) – 95.9 (10^{8}) 89.0 (10^{8}) – 91.4 (10^{8})	**Radio (Licensed)**	GSM Cellular Services
82.4 (10^{8}) – 84.9 (10^{8}) 86. 9 (10^{8}) – 89.4 (10^{8})	**Radio (Licensed)**	Cellular Telephone Service
74.7 (10^{8}) – 76.2 (10^{8}) 77.7 (10^{8}) – 79.2 (10^{8})	**Radio (Licensed)**	Ultra-high Frequency (UHF) Television
46.2 (10^{8}) – 46.7 (10^{8})	**Radio (Licensed)**	Citizens Band (CB) Radio
42.0 (10^{8}) – 45.0 (10^{8})	**Radio**	Wireless Local Loop
3.0 (10^{8}) – 30.0 (10^{8})	**Radio**	Mobile Radio
4.6 (10^{8}) – 4.9 (10^{8})	**Radio**	Cordless Telephones

WHAT IS WIRELESS NETWORKING?

Wireless networking is exactly what it sounds like: transferring data to users in a network without the use of a wired cable. In the simplest terms, the principle of transmission for wireless data networks is the same as for regular radio transmissions. In short, signals are sent to a radio transmitter linked to an antenna that converts the radio frequency signals into electromagnetic waves that travel through the air. These waves are intercepted by a receiving antenna that converts them back into radio frequency (RF) signals.

> **LAN (local area network)**
> A group of computers connected over a communications medium for the purpose of sharing access to centralized resources (files, printers, CD-ROM products).

> **Access point (AP)**
> For wireless Ethernet, a transceiver device known as an access point is connected to the wired network using a standard Cat5 cable. Most APs are about the size of a book and include a stubby antenna. The access point, or transceiver, is itself either a switch or a hub that, at a minimum, reinforces the signal and transmits the data between users on the wireless network, one or many wireless devices, and the Internet backbone (usually wired Ethernet).

> **NIC**
> A network interface card is an internal or external hardware card computer circuit board or card that allows a computer to be connected to a network. A wireless NIC translates radio signals sent via an access point to your computer, connecting it wirelessly to a network. Recently, most laptops come with wireless NICs installed.

Figure 1–4   How Wireless Works.

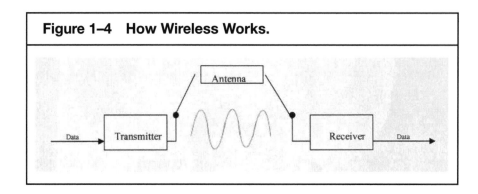

In the case of wireless **LANs**, this transmission and reception is done using **access points**. Access points (APs) are generally small boxes about the size of a book with a stubby antenna extending from them. They house a transmitter, a receiver, an antenna and a piece of equipment that acts as a bridge to your wired network. In addition, if authentication is required on the network, the AP works to encrypt and decrypt data streams (more on this in Chapter 5). On the other end, the end-user's "receiver" is the wireless **network interface card** (**NIC**) installed in his/her desktop or laptop PC or PDA (Personal Digital Assistant—Palm Pilot or similar). Increasingly, PDAs are coming with wireless adapters built in, and most laptops manufactured in the last couple of years have wireless NICs installed by default.

One access point can serve several users, but bandwidth gets divided between them, so many users can mean a slower connection. Generally, wireless network access is slower than direct wired connections to the same source, though the speed difference can vary from negligible to noticeable. The signal from an AP can transmit from one hundred to three hundred feet, but certain structural elements, like elevators or concrete stairwells, can cause interference. More information on that topic will be provided later.

Because wireless networks generally utilize the unlicensed portion of the electromagnetic spectrum (2.4–5.8 GHz range), there is typically a lot of noise within that range that has to be filtered out. Other devices, such as garage door openers, baby monitors, cordless telephones, and wireless intercom systems use the same range, so wireless devices must contain components that can pick out the correct signals amid the flotsam and jetsam of radio waves. One way in which wireless data can be transmitted with greater signal integrity is to utilize spread-spectrum technologies.

In one of those great "I didn't know that!" stories, the patent for spread-spectrum technology, granted in 1942, actually belongs to the actress Hedy Lamarr and avant-garde twentieth-century composer George Antheil. The story goes that during a dinner party, the two began discussing ways in which guided torpedoes could be used without the enemy jamming the radio waves responsible for guiding the torpedo to its target. Lamarr, who had been married to an arms dealer in her native Austria, conceived of the idea of using radio waves of differing frequencies to avoid the signal being detected and jammed. The patent was held in secret for many years, and Lamarr and Antheil's idea never saw practical application during World War II. Eventually, the military did adopt spread-spectrum technology, where it still remains a favorite means of transmitting data. Years later, the idea found new applications in wireless networking utilizing spread-spectrum technologies.

Due to the delay in use of this technology by the military, in part because of the massive mathematical calculations involved in implementation prior to the microcomputer boom, neither Lamarr nor Antheil saw royalties from this patent. However, their co-invention was honored by the Electronic Frontier Foundation (EFF) in 1997 with a Pioneer Award.

Figure 1–5   Hedy Lamarr and George Antheil, Inventors of Spread Spectrum. (Photos used with permission of their estates.)

Lamarr   Antheil

Bandwidth
In wired networks, the size of a network "pipe" or medium for communication. In wireless, it describes the transmission capacity in terms of a range of frequencies. Used to indicate how much, or how fast, data can be transmitted across a telecommunications line or network connection in a period of time, usually one second; sometimes used synonymously with data transfer rate, throughput, and line speed.

Frequency hopping
A transmission method by which a carrier "hops" packets of information (voice or data) over different frequencies, according to an algorithm.

Direct sequence
In this transmission method, the stream of information to be transmitted is divided into small pieces, each of which is allocated to a frequency channel across the spectrum. A data signal at the point of transmission is combined with a code that divides the data according to a spreading ratio. The code helps the signal resist interference and also enables the original data to be recovered if data bits are damaged during transmission. While it is more expensive and uses more power than frequency hopping spread spectrum, it is also more reliable.

Time hopping
Using what is commonly referred to as *Ultra-Wide Bandwidth* (UWB), time hopping (sometimes referred to as "digital pulse" or "impulse radio") uses extremely short pulses, or "chirps," where the signal is switched on and off in rapid succession in a digital application. The pattern on the on/off states of the frequency can be used to transmit data digitally where the early arrival of a pulsed signal represents a "0" and a late arrival represents a "1."

Spread-spectrum networks utilize wide band signals that tend to mimic spectrum noise (unwanted signals, often caused by background radiation). Because of this, spread-spectrum transmissions are very difficult to detect, intercept, and demodulate. Methods of data transmission vary, but all spread-spectrum devices utilize a *key* or *code* attached to the communication channel. This key has the effect of greatly expanding the signal **bandwidth** by inserting a higher frequency wave, diffusing the information being sent over a wider range. On the other end of the transmission (the receiving end), the process must be reversed, condensing the information back into its original bandwidth. To achieve this, the other end must have the same key as the sender. Spread-spectrum transmission schemes can be further broken down into **frequency hopping**, **direct sequence** and **time hopping** methodologies.

FREQUENCY HOPPING

The first spread-spectrum devices, like the one envisioned by Lamarr and Antheil used frequency hopping schemes to transmit data. Such devices used narrowband frequency modulated (FM) signals that rapidly jump (or hop) from one frequency to another within the band in a seemingly random fashion. In the original patent filed by Lamarr (using her married name Markey) and Antheil, the controller governing the "hopping" of frequencies somewhat resembled a player-piano roll with pins that laid out the course the signals would hop to. The sender and recipient would each have keys to the transmission pattern enabling each to decipher the data transmitted in the signal.

One rather simplified way to think about frequency hopping spread spectrum would be to envision a spread-spectrum transmission as a series of cars traveling on an expressway. (See Figure 1–6) Each car is carrying one piece of a large puzzle that will be assembled at the destination point. In a frequency hopping scheme, the cars depart out of order. In some cases, if a car runs into interference along the way, another car is sent in its place when the interference subsides.

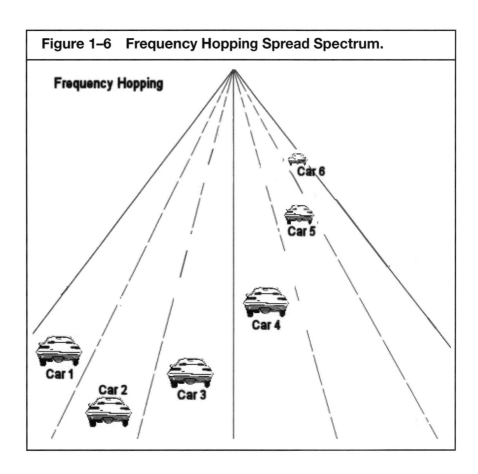

Figure 1–6   Frequency Hopping Spread Spectrum.

Bluetooth
A low-cost, short-range wireless specification (in the 2.4 GHz range) that allows for radio connections between devices within a ten meter (about thirty foot) range of each other, including laptops, APs, phones and printers. Can be used to create ad hoc wireless networks for printing, PDA downloads, etc. The Bluetooth specification was ratified into the IEEE *802.15* communications specification, which is fully compatible with Bluetooth 1.1. The name comes from tenth-century Danish King Harald Blåtand (Bluetooth), who unified Denmark and Norway.

Today, frequency hopping is a good way to get around interference in a congested, unlicensed portion of the radio spectrum. The primary technology utilizing frequency hopping schemes these days is the **Bluetooth** standard developed by Ericsson, which will be discussed later in this chapter. Unfortunately, frequency hopping devices are not compatible with other spread-spectrum technologies that use other methods, such as direct sequence or time hopping, though they are typically less expensive and require less power.

DIRECT SEQUENCE

Direct sequence spread spectrum refers to a transmission scheme in which the data is transmitted all at once, but distributed among the range of frequencies within the band. Direct sequence usually requires a very high bandwidth, expressed generally in megahertz (MHz) rather than Hertz or kilohertz like many low—frequency radio transmissions. In addition, direct sequence transmissions send the data all at once, unlike frequency hopping schemes. In a

Phase-shift key
In direct sequence spread-spectrum transmissions, the phase-shift key is a pseudo-random code generated by the sender. That code is required by the receiver to interpret the signal.

direct sequence transmission, the **phase-shift key** (a pseudo-random code generated by the sender) inserted or superimposed on to the carrier frequency has the effect of spreading out the bandwidth so that the transmission sounds like noise to anyone other than the intended recipient who has the same pseudo-random code at the other end. The way this works lies in the pseudo-random number generator that exists within each radio transceiver involved in the communication. It's called a pseudo-random number because the number is derived from a mathematical formula or algorithm and is therefore not truly random. The random number formula depends upon the seed value that starts the sequence, so when both parties involved in the transmission begin with the same seed value, their transmissions will be synchronized, allowing each side to decode the data sent by the other. There is little danger of a third-party eavesdropper happening upon the same seed value and intercepting the transmission because seed values can vary infinitely, as the seed value is a truly random number.

Let's return to our metaphor of a series of cars traveling on an expressway. Again, each car is carrying one piece of a large puzzle that will be assembled at the destination point. (See Figure 1–7) In a direct sequence scheme, all of the cars depart in an orderly fashion, beginning with car number one and concluding with the final car (the number of cars depends upon how many channels are allocated and can run into the thousands). Often, signal interference affects more than one channel, so direct sequence systems tend to lose more data than frequency hopping systems, where data can switch among different channels in a nonsequential fashion.

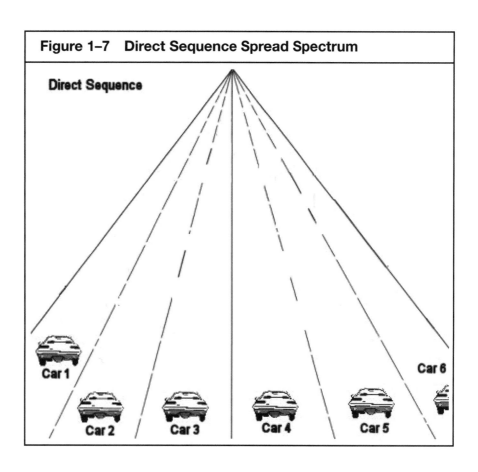

Figure 1–7   Direct Sequence Spread Spectrum

Ultra-Wide Bandwidth (UWB)
Ultrawideband, also called digital pulse, is a wireless technology for transmitting digital data over a wide swath of the radio frequency spectrum with very low power. Because of the low power requirement, it can carry signals through doors and other obstacles that tend to reflect signals at more limited bandwidths and a higher power. It can carry large amounts of data and is used for ground-penetrating radar and radio locations systems.

TIME HOPPING

Time hopping is a relatively new concept within the field of spread-spectrum technologies. Using what is commonly referred to as **Ultra-Wide Bandwidth** (UWB), time hopping (sometimes referred to as "digital pulse" or "impulse radio") uses extremely short pulses where the signal is switched on and off in rapid succession in a digital application. The pattern on the on/off states of the frequency can be used to transmit data digitally, where the early arrival of a pulsed signal represents a "0" and a late arrival represents a "1." Because of the low power requirement, UWB can carry signals through doors and other obstacles that tend to reflect signals at more limited bandwidths and a higher power. At the time of this writing, UWB had not been widely implemented except for ground-penetrating radar and radio locations systems, but its low power and low cost make it an attractive short distance networking option.

Packet-switched network
A type of network in which relatively small units of data called packets are routed through a network based on the destination address contained within each packet. Breaking communication down into packets allows the same data path to be shared among many users in the network. Most traffic over the Internet uses packet switching rather than circuit-switched networking. Voice calls using the Internet's packet-switched system are possible (VoIP, or Voice over IP, uses this). Each end of the conversation is broken down into packets that are reassembled at the other end.

FDMA (Frequency Division Multiple Access
FDMA is the division of the wireless cellular telephone band into 30 channels, each of which can carry a voice conversation or, with digital service, digital data. Used in the analog Advanced Mobile Phone Service (AMPS) phone system. With FDMA, each channel can be assigned to only one user at a time. D-AMPS (Digital-Advanced Mobile Phone Service) also uses FDMA but adds time division multiple access (TDMA) to get three channels for each FDMA channel instead of just one.

TDMA (time division multiple access)
A technology for shared medium (radio/wireless) networks, TDMA allows several users to share the same frequency by dividing it into different time slots. The users transmit in rapid succession, one after the other, each using their own timeslot. This allows multiple users to share the same radio frequency while using only the part of its bandwidth they require. This technology is used in *GSM* networks.

CDMA (code division multiple access)
According to Wikipedia.org CMDA is "a method of multiple access that does not divide up the channel by time (as in TDMA), or frequency (as in FDMA), but instead encodes data with a certain code associated with a channel and uses the constructive interference properties of the signal medium to perform the multiplexing. CDMA also refers to digital cellular telephony systems that makes use of this multiple access scheme." Phones and other devices that have a cellular radio built in use GSM/CDMA.

WIRELESS GENERATIONS

Aside from purely military applications, the use of wireless technologies for *data* transmission did not really progress until the late 1960s with the creation of the first cellular radio system. At first these systems were primarily used for voice transmission. Eventually, packet-switching technologies for data transmission were developed and perfected. However, it was not until quite recently, in the early 1990s, that wireless **packet-switched networks** began to be developed in earnest.

FIRST GENERATION WIRELESS

From the late 1970s through the 1980s, wireless networks used analog (as opposed to digital) modulation techniques. The standard used was known as Analog Mobile Phone System, or AMPS, and required the end-user phone, or terminal, to be equipped with large batteries, often toted around like a small briefcase. Compared with current systems, first generation wireless used relatively high-powered signals within the range of eight hundred MHz. The AMPS system was developed by Bell Labs in the early 1970s and is still occasionally used today.

SECOND GENERATION WIRELESS

As the technology progressed through the early 1990s, wireless networks moved largely from analog to digital. The switch to digital transmission increased carrier capacity many times over that of analog systems. These newer digital technologies used a few different spread-spectrum techniques, including **Frequency Division Multiple Access (FDMA)**, **Time Division Multiple Access (TDMA)**, and **Code Division Multiple Access (CDMA)**. However, these earlier systems, while sufficient for voice transmission, were still much too slow for Internet data transmission. It wasn't until much later that wireless technologies became viable for the requirements of large data transmissions.

THIRD GENERATION WIRELESS

According to the FCC's Web site on 3G (third generation) Wireless, "Key features of 3G systems are a high degree of commonality of design worldwide, compatibility of services, use of small pocket terminals with worldwide roaming capability, Internet and other multimedia applications, and a wide range of services and terminals." One key feature of third generation wireless is its ambition to interface with wireless LAN operations. Newer products currently in development strive to integrate large area networks with smaller wireless LANs by combining services and equipment into one device. Another feature is a push to add additional spectrum ranges for personal devices, acknowledging "the need and urgency for the United States to select radio frequency spectrum to satisfy the future needs of the citizens and businesses for mobile voice, high speed data, and Internet accessible wireless capability" (NTIA 2000). Clearly wireless and wireless-capable devices are here to stay.

TYPICAL OPERATION OF WIRELESS NETWORKS

Having a general understanding of how the typical wireless network (WLAN) operates is beneficial, whether you find yourself in the position of being the sole individual burdened with the responsibility for installing a wireless network or simply need to communicate with those who will be installing the technology (outside vendors, IT departments, etc.). A basic understanding will help you ask appropriate questions and make efficient decisions. To understand how a wireless network functions, it's helpful to compare it to how a standard wired network works.

WIRED LAN

You probably already have experience setting up or assisting in the maintenance of a wired LAN. Wired Ethernet typically uses category five (Cat5) twisted pair copper wire cabling, usually run through spaces above the ceiling and between walls, or occasionally through floor conduits. The cables running from a server to a router, switch, or hub and out to individual workstations usually terminate in a data jack located on the wall near the workstation. These jacks look very much like telephone jacks, but are slightly larger. (see Figure 1–8)

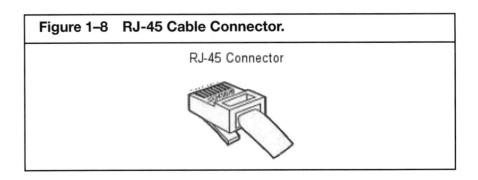

Figure 1–8   RJ-45 Cable Connector.

RJ-45 Connector

Cat5 cables can transmit signals up to ten Megabits per second (Mbps), although fast Ethernet transmits at one hundred Mbps. 10Base-T cabling for Ethernet has a distance limitation of approximately one hundred meters (328 feet) before **attenuation** makes the signal degrade too much to transmit data. To overcome long distances, Ethernet networks employ the use of switches, routers, and hubs or repeaters to amplify signal strength before passing data on to other segments of the network. (See Figure 1–9) Sometimes, Ethernet LANs employ bridges to connect the Ethernet LAN to a different type of network, such as a wireless network. Access points (APs) work as bridges between wired and wireless networks, doing the work of "translating" between the two types of networks.

Attenuation
The dissipation of the strength of a signal as it is transmitted.

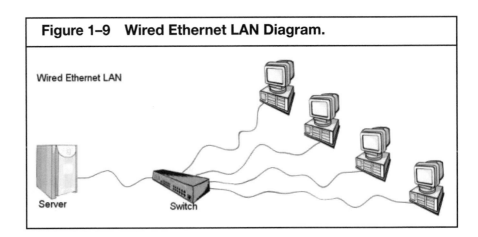

Figure 1–9   Wired Ethernet LAN Diagram.

Wired Ethernet LAN

Server   Switch

WIRELESS LANS

While it is possible to operate a stand-alone wireless network, it's more likely that you will want to use it as an extension of your wired infrastructure. For wireless Ethernet, a transceiver device known as an *access point* is connected to the wired network using a standard Cat5 cable. Transceivers are usually hidden within ceilings or closets, although the antenna is sometimes in plain view. The access point, or transceiver, is itself either a switch or a hub that, at a minimum, reinforces the signal and transmits the data between users on the wireless network and the wired Ethernet.

On the other end of the wireless network, users with laptops or PCs containing wireless network interface cards (NICs) or wireless-enabled PDAs (Palm Pilots, Treo Visors, and the like) access the wired network in a fairly seamless fashion. In some libraries, users who bring their own laptops or PDAs to the library must register their NIC before accessing the network. This setup, more frequently used in academic or special library settings, provides a measure of security for the library, as it also allows the library to keep track of the number of users accessing the wireless network.

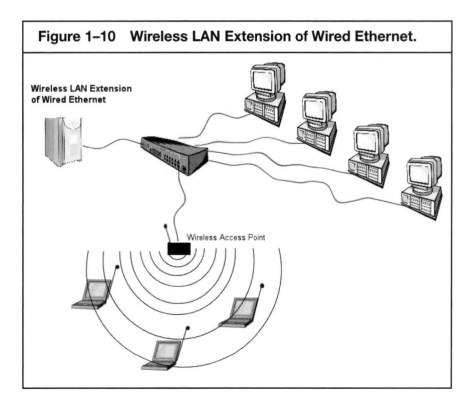

Figure 1–10   Wireless LAN Extension of Wired Ethernet.

Wireless LAN Extension of Wired Ethernet

Wireless Access Point

WPAN
A wireless personal area network is typically limited to short-range needs and is largely used to connect wireless devices in a single work area. A WPAN might be used to transfer data between a handheld device and a wireless-enabled desktop machine or printer. It could also be used to send data from wired PCs to a wireless-enabled printer in the room, which is also being used for printing by wireless devices (laptops, PDAs, smart phones). The receiving device (printer in this case) would need a wireless network card to connect. WPANs are sometimes used in homes to provide wireless connections for home security alarms, appliances, and entertain-ment systems. Typically, a WPAN uses some technology that permits commun-ication within a short range of about ten meters, or thirty feet. The primary tech-nology for WPAN is Bluetooth, from which specification the IEEE 802.15 standard was created, approved in early 2002.

WMAN
The IEEE 802.16 wireless metropolitan-area network (WMAN) standard is popularly known as WiMax. It is used in many current initiatives by municipalities to provide free or low-cost wireless broadband access to residents over a broad geographic area, sometimes known as "municipal Wi-Fi." This can include putting access points, network bridges and other equipment on water towers, rooftops, etc. throughout a city center or in areas where broadband cabling has not yet been extensively laid.

For a stand-alone network, instead of a simple access point, you would use a *wireless router* or *switch*. This houses all the functions of an access point in terms of communicating with wireless network cards and encrypting/decrypting data transfers, but also has the ability to share a DSL or cable modem Internet connection among a number of computer devices (laptops, desktop PCs, PDAs, printers, etc.), all of which are equipped with wireless NICs. This may be feasible if you are being provided Internet access for the public by a local ISP but have a different vendor for your staff wired Internet connectivity. Some public libraries have found themselves in this situation, and for security reasons find setting up a completely separate patron network to be useful. The basic principles of the wireless network as well as the planning issues we will be covering in the following chapters will apply whether you are extending an existing wired LAN into a wireless environment or setting up an entirely separate wireless network. For simplicity, we will concentrate on wireless extensions to wired LANs, as this is by far the predominant setup in libraries.

TYPES OF WIRELESS NETWORKS

Wireless technology is now widely used, most prevalently by cellular telephones. Close behind is wireless data networking, which can be transmitted in several ways for different uses. You may already have a **WPAN** (Wireless Personal Area Network) in your home, either to connect your laptop and your home printer without cables or to enable your home alarm system to send data to itself. You may get your home Internet service via a **WMAN**, or Wireless Metropolitan Area Network, also known as municipal WiFi, if your city or town government offers, perhaps in conjunction with private services, free or low-cost wireless access to the Internet for its residents. The same cellular technology that allows you to call your spouse to ask which type of ice cream to bring home can also, via **WWAN**, or Wireless Wide Area Networks, allow you to surf the Web from your cell phone while you wait in the supermarket checkout aisle. Although we will discuss each of these briefly, we will concentrate on short-range (local area) wireless networks, specifically wireless LAN, or **WLAN**, technologies for libraries.

WWAN
A wireless wide area network is a form of wireless network which differs from a WLAN (wireless local area network) in that it uses cellular network technologies such as GPRS, CDMA, GSM, or CDPD to transfer data. These cellular technologies are offered by cellular service providers for a monthly usage fee. They go beyond the basic cellular telephony and include data transfer, and since they're offered by nationwide cellular companies, can offer nationwide coverage and service. Some computers now have integrated WWAN capabilities, which means they have a cellular radio (GSM/CDMA) built in, allowing the user to send and receive data.

WLAN
Wireless local area networks use radio waves instead of a cable to connect a user device, such as a laptop computer, to a network. They provide Ethernet-type connections over the air and operate under the 802.11 family of specifications developed by the IEEE. A wireless LAN can serve as a replacement for or extension to a wired LAN.

WPAN (WIRELESS PERSONAL AREA NETWORK)

A Wireless Personal Area Network is typically limited to short-range needs and is generally used to connect wireless devices in a single work area. A WPAN might be used to transfer data between a handheld device and a wireless-enabled desktop machine or printer. It could also be used to send data from wired PCs to a wireless-enabled printer in the same room. This printer also might be used for printing by other wireless devices, such as laptops, PDAs, and smart phones. The receiving device (a printer in this case) would need a wireless network card to connect. WPANs are sometimes used in homes to provide wireless connections for home security alarms, appliances, and entertainment systems.

Typically, a wireless personal area network uses some technology that permits communication within a short range of about ten meters, or thirty feet. The primary technology for WPAN is *Bluetooth*, from which specification the IEEE *802.15* standard, approved in early 2002, was created. According to the WPAN entry on TechTarget.com:

> A key concept in WPAN technology is known as *plugging in*. In the ideal scenario, when any two WPAN-equipped devices come into close proximity (within several meters of each other) or within a few kilometers of a central server, they can communicate as if connected by a cable. Another important feature is the ability of each device to lock out other devices selectively, preventing needless interference or unauthorized access to information.
> . . . The objective is to facilitate seamless operation among home or business devices and systems. Every device in a WPAN will be able to plug in to any other device in the same WPAN, provided they are within physical range of one another. In addition, WPANs worldwide will be interconnected. Thus, for example, an archeologist onsite in Greece might use a PDA to directly access databases at the University of Minnesota in Minneapolis, and to transmit findings to that database.

HomeRF

An early WPAN application standard was HomeRF. Home Radio Frequency uses a frequency hopping standard in the 2.4 GHz range (like 802.11b) and was developed specifically for home use. This standard supported both voice and data transmission, and was designed to lessen the interference from home appliances.

HomeRF was intended as an inexpensive and simple alternative to setting up complicated 802.11b networks for those unfamiliar with computer networking who still wanted the flexibility of wireless in their homes. HomeRF did not live up to its own expectations, however, and components for these systems were quite expensive. In contrast, the prices for 802.11b parts dropped considerably, making it a more affordable solution for home networking enthusiasts as well as businesses. As of January 2003, the HomeRF Working Group, a consortium developed to promote and market HomeRF technologies was officially disbanded, marking the end of HomeRF as a viable commercial product.

Bluetooth

Like HomeRF, the Bluetooth standard, developed by Ericsson along with other technology giants such as Nokia, uses spread-spectrum frequency hopping in the 2.4 GHz range. Intended as a replacement for the short-range data cable, Bluetooth was originally conceived as a wireless peripheral interface linking printers, monitors, keyboards, and other peripheral devices to the CPU. Bluetooth was intended as a means of simplifying communications by automatically linking up to eight different devices (such as PDAs, mobile phones and headsets, televisions, and DVD players) within a ten meter range using a small radio module built into the equipment. Bluetooth is not compatible with either 802.11b or HomeRF. Though HomeRF is virtually obsolete, Bluetooth continues to be used despite being plagued by slow production and higher-than-anticipated costs. Although Bluetooth has been lagging behind in development when compared to Wi-Fi, innovative devices using the Bluetooth standard continue to appear at industry trade shows, making the ultimate future of this wireless networking standard impossible to predict, especially with the growth of PDAs and other handheld devices capable of wireless communication. You are most likely to see Bluetooth being used to connect peripheral devices wirelessly and create temporary ad hoc wireless networks for specific uses.

WMAN (WIRELESS METROPOLITAN AREA NETWORK)

One of the hottest trends in wireless right now is the municipal wi-fi movement in which towns and cities are providing free or low-cost wireless broadband access to residents over a broad geographic area. This can include putting access points, network bridges, and other equipment on such structures as water towers or rooftops throughout a city center or in areas where broadband cabling has not yet been extensively laid. WMANs can cover a few blocks, an entire neighborhood, or with the advent of **WiMax**,

WiMax (Worldwide Interoperability for Microwave Access)
Popular name for broadband wireless networks based on the IEEE 802.16 wireless metropolitan-area network (WMAN) standard. IEEE 802.16 is working group number 16 of IEEE 802, specializing in point-to-multipoint broadband wireless access, and aims to improve on 802.11 (Wi-Fi) both in performance and coverage distances. WiMAX is designed to extend local Wi-Fi networks across greater distances, such as a campus or municipality, as well as to provide last mile connectivity to an ISP or other carrier many miles away. (See also **WMAN**)

an entire municipality. WiMax is the popular name of the IEEE *802.16* wireless metropolitan-area network standard. WiMAX is designed to extend local Wi-Fi networks across greater distances, such as a campus or municipality, as well as to provide last mile connectivity to an ISP or other carrier many miles away. It is fairly new and is not, as of this writing, being heavily used. However, as the debate over municipal wi-fi continues, WiMAX options will become more relevant.

Recent challenges to these "muni Wi-Fi" citywide initiatives are of some concern. A bill was introduced in Congress in the summer of 2005, the Broadband Investment and Consumer Choice Act of 2005, designed to revise or rewrite the Telecommunications Act of 1996, which includes a section that specifically limits the ability of local governments to deploy public broadband systems. Many cities, including Philadelphia, New Orleans, and New York City, are trying to negotiate municipal wireless access for their residents, but are meeting with resistance from Internet service providers and free-trade advocates. The debate is far from over. Municipal wireless will be discussed further in Chapter 2.

WWAN (WIRELESS WIDE AREA NETWORK)

A Wireless Wide Area Network, differs from a WLAN (Wireless Local Area Network) in that it uses cellular network technologies such as GPRS, CDMA, GSM, or CDPD to transfer data. These cellular technologies are offered by cellular service providers such as Cingular Wireless, Sprint PCS, or Verizon Wireless for a monthly usage fee. They go beyond the basic cellular telephony and include data transfer, and since they're offered by nationwide cellular companies, can offer nationwide coverage and service. "Smart phones" used to surf the Web and transfer images often use WWANs. Some computers now have integrated WWAN capabilities, which means they have a cellular radio (GSM/CDMA) built in, allowing the user to send and receive data. An employee working away from his/her employer may use WWANs to access the Internet and send e-mail while away from the office. Here's a good explanation of how WWANs function in such a situation:

A portable computer with a wireless WAN modem connects to a base station on the wireless networks via radio waves. The radio tower then carries the signal to a mobile switching center, where the data is passed on to the appropriate network. Using the wireless service provider's connection to the Internet, data communications are established to an organization's existing network.

Wireless WANs use existing cellular telephone networks, so there is also the option of making voice calls over a wireless WAN. Both cellular telephones and wireless WAN PC cards have the ability to make voice calls as well as pass data traffic on wireless WAN networks. (Chaplin 2002, p 2)

The biggest strength of WWANs is security. These networks use sophisticated encryption and authentication methods, making them faster and more secure than your average wireless data network.

WLAN (WIRELESS LOCAL AREA NETWORK)

In this book, we are largely concentrating on short-range (local area) wireless networks, specifically wireless LAN (WLAN) technologies for libraries. Wireless local-area networks use radio waves instead of a cable to connect a user device, such as a laptop computer, to a network which may have other wireless components, but which nearly always has an ultimate wired connection to a network backbone. They provide Ethernet-type connections over the air and operate under the 802.11 family of specifications developed by the IEEE. In Chapter 2 we will discuss some of the reasons why a library might want to implement a WLAN, either as a replacement for, or extension to, a wired LAN.

802.11

The **802.11** standards were developed by the Institute for Electrical and Electronics Engineers (IEEE) in 1998. These standards specify an "over-the-air" interface between a wireless client and a base station, or access point, as well as among wireless clients. The 802.11 standards can be compared to the IEEE standard for Ethernet for wired LANs and use Ethernet protocols and CSMA/CA (carrier sense multiple access with collision avoidance) for path sharing. This differs from Ethernet, in that standard wired Ethernet uses CSMA/CD, or carrier sense multiple access with collision detection. For further explanation of these protocols, see the glossary. What is important to understand is that the standards for wireless 802.11 and wired Ethernet work on similar principles.

The standards continue to evolve, so that within the 802.11 standard there have emerged (and continue to emerge) a series of sub-standards designated by alphabetic characters. The **802.11b** standard, also known as **Wi-Fi**, which replaced the 802.11 standard in 1999, has been heavily used for wireless LANs, and describes a wireless networking data rate of up to eleven megabits

---

802.11

A group of wireless specifications developed by a group at the Institute of Electrical and Electronics Engineers (IEEE) in 1998. These specifications are used to manage packet traffic over a network and ensure that packets do not collide, which could result in loss of data, while traveling from device to device. As the standard 802.11 has evolved, a series of sub-standards designated by alphabetic characters has emerged.

802.11b

Also known as "Wi-Fi," devices using this standard operate in the 2.4-GHz band and with rates of up to eleven Mbps. A very commonly used frequency; microwave ovens, cordless phones, medical and scientific equipment, and Bluetooth devices all work within the 2.4-GHz ISM band. Like other 802.11 standards, 802.11b uses the Ethernet protocol and CSMA/CA (carrier sense multiple access with collision avoidance) for path sharing.

Wi-Fi

Short for "wireless fidelity," the generic term for 802.11 technology. Originally describing the 802.11b standard, it is now used generically for the overall technology. Many airports, hotels, and other locations all known as hot spots offer public access to Wi-Fi networks so people can log onto the Internet and receive e-mails on the move.

802.11a
Operating in the five-GHz frequency range with a maximum fifty-four Mbit/sec. and signaling rate, this frequency band isn't as crowded as the 2.4-GHz frequency because it offers significantly more radio channels than the 802.11b and is used by fewer applications. It has a shorter range than 802.11g, is actually newer than 802.11b and isn't compatible with 802.11b.

802.11g
This standard is similar to 802.11b but supports signaling rates of up to fifty-four Mbps. It also operates in the heavily used 2.4-GHz ISM band but uses a different radio technology to boost overall throughput, and is compatible with 802.11b, but not 802.11a.

per second (Mbps) at a frequency of 2.4 GHz. The quoted data rate is a little misleading, because "it refers to the total physical layer capacity, much of which is used by the protocol itself, so it is not actually available for data." (Dornan 2002) In reality, the actual transfer rate for 802.11b tops out at between two and four Mbps. It is also prone to interference from cordless telephones, cellular radios, and some remote controls. On the other hand, **802.11a**, adopted in November 2001, boasts a fifty-four Mbps data capacity and operates in the five GHz range. Thus it is good for large data applications such as video, and also experiences much less interference in its range than 802.11b, so often exceeds thirty Mbps actual use. However, 802.11a products work on a much shorter leash: they operate at about a one hundred foot range, compared to three hundred feet for 802.11b, so more access points are required for effective coverage. The bulk of products designed between 1999 and 2003 ended up using the 802.11b standard.

In order to satisfy the demand for higher data capacities that are compatible with older standards, IEEE has now established the **802.11g** standard. 802.11g, like 802.11b operates in the 2.4 GHz range, but its processing speed is much greater, boasting data rates of fifty-four Mbps like the 802.11a standard. But unlike 802.11a, 802.11g equipment is backward-compatible with the old 802.11b standard, basically by "ramping down" its transfer rate when it encounters an 802.11b device. This relationship, however, is one way as the Wi-Fi equipment that accommodates 802.11b cannot support 802.11g.

STOP! Before your head begins to spin (if it hasn't already) with letters, numbers, and bit rates, keep this in mind: many of the currently available access points now provide support for a, b and g standards all in one. That means you can provide wireless access to a broader range of equipment, and thus to a broader range of users. It's important to be aware of the benefits and limitations of each standard, however, if you have an unusual environment or several locations to cover. You may want to look at what will give you the best coverage in your particular space. More information on space planning is provided in Chapter 3.

Although the 802.11 specification originally called for both a frequency hopping spread-spectrum version as well as a direct sequence spread-spectrum version, the two were mutually incompatible, and the direct sequence version became the overwhelming favorite for wireless networking. However, other popular standards for wireless networking do use frequency hopping as their mode of data transmission.

Of the standards described in this chapter, it is most likely that libraries that have not already implemented wireless technologies will choose to go with the 802.11 standards. The decision is really one that has already been made. The vast majority of laptops equipped with wireless network interface cards (NICs) use the 802.11b or 802.11g standards, and these are the standards that patrons will expect when they walk in the door of the library. Many new wireless cards are able to access multiple versions of the standard for greater flexibility.

Part of your initial planning will need to be deciding, based on your library's situation, which options are right for your network and your users. In the next chapter we will discuss why you might be considering wireless, and what it can do for you.

2 WHY GO WIRELESS?

OVERVIEW

- **Why Go Wireless?**
- **Wired vs. Wireless**
- **Specific Situations That Wireless Can Serve**
- **The (Near) Future of Wireless in Libraries.**
- **The Best Reason To Go Wireless**

WHY GO WIRELESS?

While the decision to "go wireless" is by no means an automatic one, there are some definite advantages that should factor into your decision-making process. Wireless networks can be a terrific boon to library services in terms of marketing, value-added services, and increased accessibility to the Internet for patrons. There are a number of reasons you may be considering a wireless network (WLAN) for your library:

- Your college or university has decided to provide your students with wireless-enabled laptops, or laptop purchase options, setting expectations that students will access the Internet and/or library resources wirelessly at the library. In addition, the administration may be considering implementing wireless throughout the campus, and you want to be sure all areas of the library will be covered and that it will be useful for your staff and patrons.
- Your municipality has decided to launch a "Muni Wi-Fi" program, providing free wireless access to a huge number of residents via wi-fi transmitters on towers, etc. There is an expectation that you will take advantage of this option for library patrons, but you're either not sure what impact it will have on your services, or you do know the impact and want to be sure the city "does it right."
- You need to provide additional access to the Internet for patrons in the building, due to increased demand and/or to meet some minimum standards set by state or grant programs, but you cannot afford to buy additional PCs or to run significantly more cabling in your building.
- You need to provide additional access to the Internet for

patrons in the building, but you have an older and/or historic building, which presents barriers to extending cabling in the structure.

- Alternately, you may have an architect's dream library, which means you have huge, open spaces which make it difficult to run more cables without exposing them to view, or exceeding useful cabling lengths.
- You want to keep up with the latest affordable technology available to keep your community or organization well placed on the technology curve.
- A library board member, departmental chair, or faculty dean has asked the question, "Why don't we have wireless? Borders has it!" This can cause the most stalwart library administrator to tear his/her hair out, but making it happen soon becomes your problem.
- Last, but by no means least—in fact possibly the most important reason—is that your patrons have repeatedly and increasingly been asking for it.

WIRED VS. WIRELESS

Chances are good that you have a *wired* local area network (LAN) in your building, providing access to the Internet and other networked resources for staff and/or patrons. In the previous chapter, we discussed the basic functions of wired and wireless LANs. Let's stop a moment and look at the relative advantages and disadvantages of wired and wireless networks.

Figure 2–1 Wired vs. Wireless Networks.	
Wired (Traditional) Network	**Wireless Network** (or Wireless Extension to a Wired Network)
Advantages: • Speed—throughput is fairly consistent. • Reliability and stability. • Security—you have control of firewalls, desktop lockdowns, etc. • Control of equipment—any equipment attached to this network is under your control.	**Advantages:** • Flexibility—formerly inaccessible areas of your library can now be used for network access. • Patrons can use their own equipment—this frees you from buying additional computers to help with demand, thus reducing the cost for each "seat" you're now adding. • Takes a load off your public computers, freeing them for other uses (word processing, databases, growing demand, etc.). • Many of the traditional security features • (firewall, anti-virus, etc.) can be implemented here, too. • Add Internet access to a new space (building extension) without expensive cabling. • Less cabling needed—great for older or historic buildings, where adding cabling can cause major structural problems.
Disadvantages: • Lots and lots of messy and expensive cabling. • If you want to expand, you often have to run new cabling, as well as buy new computers to attach to the network. • Your security setup is unlikely to allow users to use their own equipment to access your network. • Patrons are "stuck" at the established wired locations, usually clustered in one or a few areas of your building(s).	**Disadvantages:** • Interference from structural elements, such as elevator shafts, stairwells, etc. can cause dead zones in your wireless coverage. • Requires wireless NIC in users' equipment to access. • Slower, lower throughput (speed) than wired. • Concerns about hijacking of bandwidth by users. • No control of equipment linked to network (users' laptops).

Wireless networks offer libraries a degree of mobility and flexibility that simply cannot be achieved with wired networks. Indeed, anyone who has taken a trip to a server room or router closet has seen the tangled mass of cabling that accompanies even a modest wired LAN. Wireless LANs, on the other hand, are generally less complicated and frequently cheaper in the long run, once you eliminate the costs of extensive cable installation, additional workstations, and maintenance and troubleshooting of wired network equipment. Let's look at some reasons wireless may be the right choice for your library:

SPECIFIC SITUATIONS THAT WIRELESS CAN SERVE

At times, wireless may be the only feasible networking option for certain buildings or spaces. Buildings containing asbestos or those designated as historical landmarks often involve a wide set of regulations pertaining to the preservation and maintenance of interior and exterior spaces. At the other end of the architectural timeline, more recent structures often contain large open spaces and very high ceilings. This represents another sort of obstacle to wired networks.

CONNECTIVITY FOR WIDE OPEN SPACES

In buildings where the architecture includes expansive open spaces, it can be rather difficult to provide traditional wired network access. Very high ceilings combined with the absence of sub floor conduits due to floors made of marble or other stone materials become barriers to the laying of network cable. The length of the wire becomes an issue in the case of very high ceilings with respect to the possibility for **attenuation** difficulties, depending upon how far away the hub or switch is. The logistics of finding suitable and convenient locations for data and electrical outlets in open areas is also a nightmare. Strewing cable across floors to connect devices is a potentially dangerous option, as you run the risk of patrons or staff tripping on stray wires. Cable lying across floors is also aesthetically unpleasant, even if concealed underneath a cable channel running over the floor.

Although open spaces are often intended as atrium or sitting areas, rather than work spaces, many people prefer to work in these pleasant locations. Laptops have made this possible, but how can we provide network access to these users for maximum productivity? Demands for more efficient, usable work spaces as well as the desire for a more mobile and flexible arrangement for furniture have prompted many libraries to consider converting open spaces to work areas. In addition, libraries that are bursting at the seams, with no new expansion in sight, will want to utilize every inch of space. Wireless networking makes this possible.

MEETING MINIMUM STANDARDS IN THE SMALL LIBRARY

Many libraries have statewide or regional initiatives that require them to meet certain minimum standards for funding, which can include having X number of Internet "connections" per capita served. For very small libraries, and some mid-size libraries in growing communities, this can be a challenge. Wireless "connections" can count toward these standards:

> The process [of implementing wireless in our library] was delayed due to a variety of reasons, but we really got going when the new state standards came out. Libraries in our size category were required to have a minimum of five public Internet stations. We only had four—and didn't have the capability to add wiring for any additional computers. The wireless option allowed us to meet the minimum standards—and look fairly "hip" and "with it" to our public!
>
> As an aside, we use the laptop in-house all the time. It is our fifth Internet station. It isn't connected to a printer and doesn't have any other programs on it, so it truly is just an Internet station. Some people prefer using the laptop and ask for it; others may start out using it and move to one of the stationary computers once there is one available. Not surprisingly, teenagers love using it because they can go anywhere in the building (it's a small building!) and sit together while they play games.
> Joyce Godwin
> Library Director
> Indianola Public Library
> Indianola, Iowa

Though wireless provides some options for user connectivity, it can also present some unique challenges for very small libraries with limited space. We had this interesting response from the director of the small library Louise frequents when she is on vacation:

> We are very small and just added wireless this summer, which everyone loves and is going smoothly. Two comments: first of all, it seemed like people with laptops were there within about ten minutes of our installing the system! How did they know that?!!! Secondly, the biggest change for us is that our seating was always adequate before, but now it's very tight at times. That is something that I hadn't anticipated. We are a resort community, so things are back to normal now that school is starting.
> Gay Anderson
> Director
> Helena Township Public Library
> Alden, MI

We got responses from several small libraries that are in resort towns like Alden, Michigan, and Arnolds Park, Iowa, where access requirements "during the season" were exponentially higher than off-season, with "temporary residents" requiring Web access to check their e-mail or verify flight reservations, etc. Wireless was an inexpensive option for them to provide the minimal Internet-only access desired by tourists, raising the profile of their town as a desirable destination at the same time.

> We at Arnolds Park like our wireless Internet. It's very fast. In the summer when we have throngs of tourists, etc., it makes the wait for a computer so much shorter since lots of people bring their own. Mediacom [broadband cable company] gives us the account for free. When I think back to the old days —I can remember so many problems with dialup connections going down and being slow and unreliable. We have very little trouble now.
> Susan Sup
> Library Director
> Arnolds Park Library
> Arnolds Park, Iowa

PROVIDING CONNECTIVITY FOR HISTORIC, OLDER AND/OR UNWIRE-ABLE BUILDINGS

A wireless network, either as an extension of an existing wired network, or as a stand-alone wireless network, may be the most viable option in buildings protected for historical reasons. Wireless networking is also ideal in structures primarily made of stone, or ones containing asbestos, which prevents drilling for cable installation or removing tiles that may contain asbestos. Single-story

Plenum
The area above a room where heating and air conditioning run. Any device placed in the plenum area must meet the UL2043 standard for fire safety.

buildings on concrete slabs and multi-story buildings with no **plenums** (space between the floor and the ceiling below) are excellent candidates for wireless.

Although it should be stressed that wireless networks work best as extensions to existing wired Ethernet networks, stand-alone wireless networks may be the only viable option in some of the situations just mentioned. At some point, there must be a connection to a service provider or network backbone, but a wireless network can dramatically minimize the need to drill holes and run expensive cabling.

Presumably you know whether your library is listed as an historic building. Chances are you've run into this every time you've tried to make modifications or additions to your "charming" Carnegie or similar building. Nonetheless, your users are demanding additional Internet access. Wireless may very well be the answer for you. Not only can you avoid drilling holes, but you may end up extending the useable field to include that equally charming patio in front of your building, facing the town square. Encouraging laptop users to work out-of-doors on nice days may help your inevitable congestion problems, too!

Whether or not your building has been designated an historical landmark, you may still have to deal with the presence of asbestos-containing materials that may be in your building. Building materials containing asbestos were used frequently from the 1940s until the early 1970s. Any buildings dating from that period that have not been renovated or been treated specifically for asbestos removal are likely to have some asbestos.

The presence of asbestos is not a problem until it is disturbed, releasing small asbestos fibers into the air. Some of the more common products that contain asbestos include pipe insulation, drywall and joint compound, textured paints, cement board, electrical wiring and cable, ceiling sprays (used for texture), ceiling and flooring tiles, mastics, and stucco. Drilling through walls, ceilings, or floors that were either coated with asbestos paints, texturizing sprays, adhesives, or tiles could conceivably release dangerous amounts of the fiber into the air, endangering both the person installing the cabling as well as occupants of the building.

OSHA standards require that building and facility owners determine the presence, location, and quantity of asbestos within a building and inform building occupants, employees, and contractors of the danger, but if there is any question at all about the possibility of the presence of asbestos-containing materials within a building, it is always wise to make inquiries and leave everything alone until the questions are answered.

FILLING THE NEED FOR FLEXIBILITY/TEMPORARY NETWORKING

One of the most important advantages of wireless networking is flexibility. Libraries that implement wireless can be more creative with the services they provide. Many libraries with wireless networks have found that they are now able to meet the needs of their patrons and staff much more effectively. Some, like the Douglas County Library System in southern Oregon, have established flexible training lab space after funding for dedicated workstations and space fell through. In another example, the Milwaukee Area Technical College revamped their bibliographic instruction program and overcame space constraints by purchasing twenty laptops equipped with wireless network cards, transportable on a cart. This "mobile classroom" allowed the library to partner with groups on campus, and eventually improved the library's image as a technology leader for the campus. Having the ability to construct computer labs on-the-fly is not only a boon to the library in terms of enhancing the bibliographic instruction program. It also demonstrates to patrons and administrators that the library is resourceful and creative in finding solutions to problems.

OFFERING POINT-OF-NEED SERVICES

Barbara Ginzburg, reporting on the uses of wireless networking at the University of Kansas Law Library, relates the factors that eventually drove the library to seek a wireless solution to their circulation and inventory needs:

> We decided that because our faculty members use so many of the library books for research, it would be more disruptive to remove books from their offices to process them than to do the inventory in their offices.
>
> When we originally began manually processing books in faculty offices and student carrels, we wrote down call numbers, titles, and bar code numbers. We then took these lists to the circulation department and checked each item out. But with 55 faculty offices, 65 carrels, and over 2,000 books to process, this task became very tedious. We decided that because we could not bring the books to the circulation desk, we had to find a way to take the circulation software to the books. We needed to be able to perform three main operations: Check out books that had item records in the system, create item records for books that had bibliographic records but no item records, and create bibliographic records on-the-fly for

items that weren't in the catalog at all. So, we'd need our circulation software, our cataloging software, and possibly OCLC available as we processed books on site.

. . . After coming to this conclusion, we decided we had three options: 1) purchase a remote bar code scanner, 2) plug a laptop with the appropriate software into the network jack in each office as we processed books, or, 3) purchase a wireless network starter pack. . . . Needless to say, we chose the third option. (Ginzberg 2001, 41–42)

While conducting an inventory, the librarian simply plugs the wireless access point into the closest network jack and proceeds to maneuver a laptop situated on a book truck around the floor, scanning the barcodes for each book. In a similar scenario, libraries that experience rushes at the circulation desk or require an extra circulation desk during special events could plug in an access point connected securely to the ILS system and create a temporary impromptu check-out desk. This extension of service represents a flexible response to customer service issues.

While wireless can offer many solutions for staff needs, you should keep in mind that wireless can be significantly slower, having a possible negative effect on productivity. This narrative from Thomas Edelblute, Public Access Systems Coordinator of the Anaheim (CA) Public Library, is a good example:

> The Anaheim Public Library believed that it would be preferable to purchase a single access point to serve staff needs than pay the cost of cabling down from the second floor to the first.
>
> The construction of the building made the cabling path difficult. We installed this in 2001. . . . Upon the introduction of the wireless computers, the staff searching the Integrated Library system complained that it was slower than the other computers. The complaining never ceased, and in 2004, when the access point became unreliable and kept going down multiple times during the day, it was decided to bite the bullet and cable the desk. Staff has been so thankful that their book catalog searching is much faster now.

In this case, wireless was a good short-term solution until a wired option was available, but was not feasible for the long term. Wireless for staff use may still be a good solution for your library, but your planning may need to include contingencies or backups.

CAMPUS-WIDE WIRELESS AND THE ACADEMIC LIBRARY

Most medium to large-sized colleges and universities have either implemented or are planning a campus-wide wireless network. According to Bill Drew, owner of the Web site *Wireless Libraries*, nearly every state has at least one public institution offering wireless services. Richard Boss states that "at least one-fourth of public libraries and two-thirds of academic libraries had a wireless LAN in the fourth quarter of 2004." (Boss 2004)

Carnegie Mellon University was among the first to install a campus-wide wireless network. Known as "Wireless Andrew," the infrastructure for the wireless network was established at the school in 1994 as a research support network. This original wireless network, which linked Carnegie Mellon's on-campus wired computing system (known as Andrew) to researchers on campus and throughout the Pittsburgh area, was made possible through a grant from the National Science Foundation and Bell Atlantic NYNEX Mobile Systems along with Lucent Technologies. The CMU project was (and still is) one of the more ambitious wireless campus networks comprised of several "niche" wireless networks.

The ease with which wireless networks can interface with each other and overlap is something that the University of South Florida considered when planning their campus-wide wireless network. This model has proven quite feasible and should encourage academic libraries considering a wireless network of their own to take the plunge regardless of the existence or nonexistence of a campus-wide network. The pieces can always be put together later.

> We took our time figuring out how we would address the wireless question. Though I am now a programmer/developer in the library, at the time all this was shaking out I was a sysadmin for the University. The library did not lead any of the wireless efforts, and my involvement as a sysadmin mostly annoyed the networking folks who wanted to lock everything down and use incompatible hardware.
>
> The problem, of course, is that vendors are well prepared to deliver solutions that appear secure for office environments, but public service environments change everything. Patrons use a broad range of equipment, most of it incompatible with the solutions the network folks were pushing. It took a while to get them to look in other markets–the hospitality industry, for example–for wireless implementations that worked for our patron base.
>
> Casey Bisson
> Library Information Technologist
> Lamson Library
> Plymouth State University
> Plymouth, New Hampshire

When the entire campus opts for wireless access, the library must attempt to include itself in the planning and implementation phases. The library is a logical place in which to test a wireless network and provides an ideal testing environment. As a location with a high service profile, its primary function is to serve students and faculty, primarily through electronic means. Students using the library often ask about wireless networking and seldom have to be informed of its existence and availability.

PUBLIC LIBRARIES, COMMUNITY COLLEGES AND THE MUNICIPAL WI-FI MOVEMENT

Public libraries, and even community colleges, may want to take advantage of municipal wi-fi as a low-cost means of adding wireless for their users. Instead of sectioning off your own network, work with the municipality to make sure you have strong coverage from their network in your library building(s), and let the patrons use "muni Wi-Fi" at little cost to you! You give up both control and some ability to troubleshoot for patrons, so be sure that if you decide to go this route, you get in on the ground floor of any such initiative. Input from a service-oriented entity like a library can be invaluable to a muni wi-fi initiative, because librarians are aware of the sorts of questions citizens may have about it based on what they've already been asking.

Which brings us to a caveat: if you feel the municipality's wireless service will not provide the speed, coverage, or access you feel your patrons should have, stand firm in demanding a separate WLAN for your library. For instance, some muni wi-fi initiatives are providing low-cost Internet access rather than free access, so if you already provide free access at your wired stations, you may feel it is not commensurate with your library's mission to provide wireless access to a for-fee municipal service without providing a free wireless option also. You know what is best for your library—if this is the only way your patrons will get wireless in the library, then by all means jump on the municipal wagon and make sure the city does it right!

One IT director of a suburban city government, who described his experience with "muni Wi-Fi," included cautionary tales of what not to do, as well as some innovative solutions:

The IT director described a pretty typical start to the municipal wireless process: senior management in the city realized they had an area of town that needed redevelopment, and someone mentioned that one way to revitalize the area and help local businesses was to provide free wireless Internet. They realized it was

Captive portal
A method of "capturing" your Web traffic and forcing users to "click-through" a user agreement, authenticate, and/or pay for service before they can access the Internet at large. If you've ever used the Internet service at a Starbucks or a Borders store, you have encountered a captive portal. A captive portal forces an HTTP client on a network to see a special Web page before being released out to the Internet. This is done by intercepting all HTTP traffic, regardless of address, until the user is allowed to exit the portal. Usually captive portals are used for authentication.

not something the city could do on its own so they looked for corporate "sponsors" to help defray costs and support issues. A city leader, who owned a tech company, offered his help.

At that point the project entered what the IT director described as "scope creep." This is where a project with modest limits grows out of proportion to the planning being done. Some less technically aware members of the planning team didn't understand that the access points they bought for home for fifty dollars were not appropriate for this project, but instead carrier grade APs would be needed, at a cost of about five thousand dollars each. Once that was understood, it became clear that to cover the area they were looking to serve would require additional bridges and repeaters to be sure to reach into all the businesses along the coverage area—a small business district.

The technical problems were many, and will be discussed further in later chapters, but one aspect that was not discussed properly until the IT director was brought on was how the target audience was actually going to *use* the service. In particular, how would they be authenticated to it? Would they be required to give information to track metrics? At this point in the process the expertise of service-oriented organizations, like libraries, would have been useful. The IT director had experience with, as it happens, *library* wireless projects and therefore was familiar with options to cope with these issues. To facilitate authentication and metric tracking with minimal fuss, the director chose to use a **captive portal** product, which allowed him to choose, via a Web interface, what information users would be required to provide before being authenticated. Once the cookie is set, users can come back repeatedly without re-authenticating until it is cleared—a simple solution used in many libraries, and a step that might have been largely forgotten until well into the project.

Challenges to Municipal Wireless

Recent challenges to these citywide wireless initiatives are of some concern. A bill introduced in Congress, the Broadband Investment and Consumer Choice Act of 2005, was designed to revise or rewrite the Telecommunications Act of 1996, and included a section that specifically limited local governments' ability to deploy public broadband systems. In an excellent article for O'Reilly Publishing's Policy Center, Richard Koman writes: "The bill says local governments that want to build a public network must issue a Request for Proposal–*through a third-party agency*–and that in the case of competing bids between private and public sector parties, the neutral agency shall give preference to private sector companies . . . ." (Koman 2005)

In this article, Koman proceeds to highlight some disturbing developments such as when "Verizon pressured the Pennsylvania State House to pass a law banning municipalities from directly offering fee-based service." And when Rep. Pete Sessions (R-Texas) introduced a bill and Florida Governor Jeb Bush passed a law banning municipalities from providing wireless services if a private corporation already offers a similar service.

Senators John McCain (R-Arizona) and Frank Lautenberg (D-New Jersey) later that year introduced the Community Broadband Act of 2005 that supports a municipality's right to provide "advanced telecommunications capability." McCain asserted, "In many rural towns, the local government's high-speed Internet offering may be its citizens' only option to access the World Wide Web. Despite this situation, a few incumbent providers of traditional telecommunications services have attempted to stop local government deployment of community high-speed Internet services. The bill would do nothing to limit their ability to compete. In fact, the bill would provide them an incentive to enter more rural areas and deploy services in partnership with local governments." This conversation is far from over, as municipalities, and libraries, attempt to help their users bridge the digital divide.

THE (NEAR) FUTURE OF WIRELESS IN LIBRARIES

In the near future, the use of wireless in libraries may result in less reliance on handwritten call numbers. Patrons can more easily save the full bibliographic record of an item, cutting down on those desperate phone calls for missing information the night before the paper is due. Users can more easily take copious notes from rare books and other non-circulating items. A user in a remote area of the library could use a virtual chat service to ask a reference question without abandoning their work area—or, better yet, the reference librarian could move around the library, providing on-the-spot service, instead of being chained to a desk and wired PC. Chat services could include VoIP (voiceover IP) as well as text-based interactions. Users could access other forms of online instruction such as tutorials and pathfinders from any space in

RFID (radio frequency identification)
A wireless data collection technology that uses electronic tags for storing data. Like bar codes, they are used to identify items. Unlike bar codes, which must be brought close to the scanner for reading, RFID uses low-powered radio transmitters to read data stored in the transponder (tag) at distances ranging from one inch to one hundred feet.

the library covered by wireless connectivity. Imagination is the key to exploiting wireless technology. Libraries have been experimenting with the use of wireless to assist in inventory and weeding projects. One intriguing development is the use of microchip-based **RFID** tags which can be accessed via wireless. A wave of a wireless wand can provide massive amounts of data to upload to a laptop being used for inventory. If costs come down, the use of these tools could end the tedious scanning of barcodes.

The Oulu University Library in Finland is using wireless PDAs to provide map-based directions to books and other material in the library itself. From an article in *Directions Magazine* in June 2003:

> A wireless PDA–based system called SmartLibrary helps users to find books and other material from the library collections. The help is provided in the form of map-based guidance to the target bookshelf on a PDA. Ekahau Positioning Engine™ software is used by the SmartLibrary system to pinpoint the accurate mobile client location.
>
> The guidance is integrated to the online catalog of the library, so that books retrieved from the catalog can be located. Wireless connectivity is provided in the form of Wi-Fi (IEEE 802.11b) network. The guidance is based on dynamic Wi-Fi positioning of the user and static location information of books. The service is a completely software-based solution, which can be provisioned atop a Wi-Fi installed for wireless Internet access, without any additional hardware.
>
> User evaluation with real library users showed that SmartLibrary saves time and makes book finding easier.

THE BEST REASON TO GO WIRELESS

The final reason we listed at the beginning of this chapter is to add wireless access because your patrons have been asking for it. You may already have noticed, as we have, that requests at your library for wireless access are becoming more and more frequent. Being able to accommodate the demands of customers is both gratifying for those of us who provide services and a smart public relations move. Patrons of all types of libraries as well as library staff and librarians are finding that the increased flexibility of wireless can bring about a change in the way library spaces are perceived. Areas that were previously thought of as "dead areas," like atriums and lobbies, or small areas such as vestibules, where desks and traditional computer workstations are not an option, suddenly become possible workspaces. Many patrons like to seek out small quiet places to work and study. In the past, such places were usually not equipped with network access, so patrons' task options were limited. However, with a laptop and wireless network card, patrons and library staff alike can turn almost any space into a multi-task workspace.

Wireless allows users to work from anywhere, sometimes even outside the building. Patrons who bring their own wireless-equipped laptops to the library can take advantage of network access while locally creating, storing, and accessing their own private data stored on their own machines. Many libraries do not offer Internet-accessible workstations outfitted with other software applications, such as word processing, spreadsheets, or database managers. Even fewer allow patrons to download files from e-mail and store them on library computers. But patrons using their own equipment, configured to take advantage of a library's wireless network, have none of these worries. Because the prices for wireless networking components have dropped considerably in recent years, implementing wireless access within the library has become a very feasible option.

At Maryellen's library, a large academic institution in an urban university setting, the first two weeks of each semester brings a flood of students to the library, all needing Internet access to register for classes, drop and add other classes, and sign on to their course sites through the university's Web portal. Invariably during this time, all of the wired network workstations are completely occupied all day long. A proposal to use the library's wireless network and wireless-equipped laptops as an ad hoc series of computer workstations would alleviate much of the stress of students waiting for online access during this crucial period of the first week of classes.

In Louise's library, a medium-sized public library in a fast-growing, affluent, technologically eager suburban area, the thirty or so public Internet access PCs are in almost constant use. The general affluence and attendant technological sophistication of the area's residents means that many patrons arrive at the library with their own laptops in hand. The area boasts a number of colleges, universities, and a medical school, so college and medical students were the first to ask if the library had wireless Internet access, followed quickly by the traveling businesspeople who haunt the library for a few weeks at a time while they consult with local

Fortune 500 companies, conducting much of their business from the library via the Internet. It became clear that wireless access would not only be used, but would ease up traffic on the public access PCs at the same time. Within literally minutes of going "live" the network was being used by patrons.

You know that adding this service will be a huge customer service boost, but don't forget that it will also be a huge public relations bonus. Don't miss the opportunity to "get the word out" to the public/campus, and boost the library's profile in the process! The next chapter will provide some ideas for *marketing* your new wireless service.

3 PLANNING FOR WIRELESS NETWORKS

OVERVIEW

- **Planning Basics**
  - o Delegating Responsibility and Establishing Communication
  - o Conducting a Needs Assessment and Answering Essential Questions
  - o Creating a Timeline
  - o Creating a Site Survey
  - o Reviewing Ranges, Bandwidth Requirements, Equipment Options, and Budget
- **Implementing Wireless Policy**
- **Marketing Your Wireless Service**

PLANNING BASICS

Okay, you've made the decision to add wireless access to your library. Now what? Although installing a Wi-Fi network is generally quick, easy and takes relatively little in terms of resources, careful planning can streamline the process and, by making sure you have what you need up front, potentially reduce initial costs and later maintenance and security headaches.

So how do you begin? Like the old journalism maxim, technology project management comes down to "Who, What, When, Where, and How." We're going to use this as our rough framework to discuss the planning stage, followed by some additional issues we think you should keep in mind as part of your larger project.

DELEGATING RESPONSIBILITY AND ESTABLISHING COMMUNICATION (WHO)

Are you solely responsible for "making this wireless thing happen," or are you working with a committee or technical department? After the planning stage, can the actual work of installing be done by a campus or city IT department? Can you work with an outside IT vendor to install the equipment and troubleshoot the network?

For very small libraries with very small budgets, it may be that one person (you!) is responsible for planning the project, as well as implementing it. You may have to think creatively about getting help, if you need it. We've heard of a few instances at small libraries where a member of the public or a board member offered their services, and often equipment, to the library to get it jump-started with wireless.

We are a small community library. A library patron came in one day and asked if we had wireless, and I said no. He is a missionary in Mexico and elsewhere, so his home base in Iowa was just for a few months a year. As a result, he relies very heavily on his laptop and has no regular Internet service at home. He came into the library and saw that all of our regular computers were busy. When I said we didn't have wireless, he said he had an access point and the equipment to set it up. He could then come to the library or the parking lot when he needed access. I'd had maybe three other patrons in the past who'd expressed interest in the service, so I said great.

We have a person who helps with our tech support at the library. We later realized it would be a security problem for our network the way the patron had it set up. I've turned the service off for the time being, and we're working on how to make that available again in the future, more securely. At the last ILA [Iowa Library Association] conference, I heard a speech on wireless networks, and the speaker's recommendation seemed to be don't do it unless you have the tech support to secure the systems.
Karen Burkett
Director
Bondurant Community Library
Bondurant, Iowa

Some libraries are able to "make a deal" with a local ISP to provide its Internet service (as a whole, or just for the wireless component—the latter requires the wireless be separated out) in return for public acknowledgment: "Internet service brought to you by Acme ISP." The Arnolds Park Library in Iowa gets free Internet service from the local cable company in return for, essentially, public goodwill for providing the service to the community. In either case, be sure you make clear with your benefactor, just as you would with a contracted company, the goals, costs and timeline of the project, as well as expected support, so that bad feelings, bad public relations, or, even worse, maintenance problems are avoided.

Establish Communication

If you decide to work with an outside vendor, or if you are working with a team or committee, be sure to establish at the start what your means of communication will be. Nothing sidelines a technology project faster, or makes it cost more, than poor communication. Decide up front how often you will meet to discuss and monitor the progress of the project. The larger the project group, the greater the likelihood of miscommunication, so it is imperative that the project group not only meet regularly, but discuss each component of the project to ensure that all members understand the project scope and their role(s).

We strongly suggest that you create an e-mail distribution list of all of the project team members to keep each other informed, whether you plan to meet regularly or not. Whatever communication method is chosen, be sure that all members of the project are aware of even the most minute changes, accomplishments or contingencies that may arise.

CONDUCTING A NEEDS ASSESSMENTS AND ANSWERING ESSENTIAL QUESTIONS (WHAT)

What are you trying to accomplish? Are you simply expanding your existing Internet access to include wireless access, or are you creating a separate, stand-alone network? Will this be for patrons only, or also for staff? Will you need to keep those functions separate? Though you may have answered these questions when going through the decision process described in Chapter 2, you may still want to conduct a brief *needs assessment* to be sure you've covered all the possibilities for your wireless access.

Conducting a Needs Assessment

Depending on the size of your organization, you may either want to do your needs assessment informally or more systematically. In a small library, or working with a small, dedicated implementation group, you can brainstorm to identify the perceived needs. Is it simply a matter of making sure that laptop-carrying patrons can use the Internet from "the comfy chairs?" Or does your staff (or director or board) have visions of doing work-related tasks from around the building? Do you conduct classes? Would you like to have the option of offering those classes in additional locations around the building by creating a "wireless laboratory?" By getting your staff involved and asking about what they want from wireless, you may be pleasantly surprised at the creative uses they propose*!*

If the intent of your wireless network is to serve only staff, then you may be able to stick with a small-scale project, particularly if the staff using the wireless network will remain in one place. While that seems contrary to the purposes of Wi-Fi, remember that its use can extend into places that would otherwise be unreachable by standard wired networks, so converting just a room or two into new classrooms or workspaces using wireless networking is a reasonable application wireless can serve. If the area is relatively small and the number of staff needing access is relatively few, then a wireless network is a feasible and inexpensive alternative to pulling new cable and installing new data outlets.

On the other hand, you may need a wireless presence throughout the library for staff to conduct inventories, collection management and development activities, or for staff to create ad hoc classrooms or training areas in various locations. Some of these options will entail a bit more planning on your part, as well as some creative thinking. Including a needs assessment document in your project planning data will not only help you to identify needs, but also will create a tangible record of the project planning group's thought process, helping you to justify costs or the necessity of the project itself.

In a larger organization you may want to conduct staff (and possibly patron) surveys about what they want to see from a wireless initiative. A point of clarification: A survey of respondents in your organization, as described, should not be confused with the *site survey*, to be discussed later in this chapter. The site survey involves mapping out your library to determine what technology will be needed where. Because many people have only the vaguest understanding of wireless access is and what it can do, you may need to include examples of possible uses in your survey. There are several good resources on how to conduct surveys, constructing survey questions, etc. We would only recommend such a time-consuming step if your project is exceptionally large, or if you really feel you need this feedback to move forward or to convince a reluctant board or oversight committee.

Needs assessments can be created from a variety of sources, from a formal survey to focus group discussions or simple observation (e.g., you've got patrons wandering around with laptops and no way to connect them to the Internet!). Most needs assessments have several basic components which include:

- identifying the **specific problem(s) you wish to address** (e.g. lack of available wired terminals, need for wandering reference services, wireless inventory needs as a result of an RFID project),
- identifying your **population to be served** (staff, public, campus),

- identifying **resources at your disposal** (staff, budget, equipment), and
- establishing **solutions to address the problem.**

From your needs assessment you can create a few basic project goals, followed by a simple timeline (see p. 47). Once you have conducted your surveys, focus groups, and/or made your observations, it's a good idea to construct a simple document outlining your needs assessment. Figure 3–1 shows an example of the sort of needs assessment document you may wish to include in your project planning.

Figure 3–1　Sample Needs Assessment Document.

Stated Need	Problem	Recommended Action/Solution
1. Extra workstations for patrons using the Internet	No available data jacks for additional wired network workstations. Cannot afford to run additional Ethernet cable. Areas of library not currently wired for Ethernet are not conducive to laying cabling.	Install wireless access points to cover unwired areas of the library. Purchase ten additional laptops configured with wireless network adapter cards for patrons to use when wired workstations are full.
2. Mobile means of conducting library inventory.	Conducting a library inventory is very time-consuming because all books have to be removed from the shelf and taken to circulation to be scanned.	Purchase and install wireless access points that will cover stack areas of the library. Purchase wireless laptop and barcode scanner that will allow library staff members to scan books in the stacks.
3. Reconfiguring of library space to provide extra staff work areas.	With the added number of staff members in the library, there are no longer enough staff computer workstations to accommodate everyone. Additional work areas are available, but they are not equipped with network connections.	Purchase and install wireless access points that will cover the staff work areas not currently equipped with network access. Purchase workstations or laptops equipped with wireless network cards that will allow staff members in unwired areas to access network resources.

An assessment document can be as long or as short as your individual circumstances dictate. What's most important is that you have a concrete record of the planning process that you can refer back to and/or alter if needed. Documenting the recognized needs, the tasks that need to be performed, and the individuals who will take on the responsibility of carrying those tasks out will assist you in maintaining a smooth flow of information between those working on the project and any stakeholders in its successful completion.

Answering Essential Questions

Jim Rosaschi, Manager of Technical Services at Sonoma County Library in Santa Rosa, California, poses a number of good questions, that a library should consider prior to implementing a wireless service:

Will you require authentication of any kind? If you authenticate by requiring a library card number, is the transmission of that library card number to your authentication server encrypted?

We will discuss security and authentication methods in Chapter 5. One common method of securing a wireless network is to encrypt the signal and require users to provide an encryption key to use the service. This key can be handed out manually by library staff, with or without registration by the user. More sophisticated systems can interface with your ILS (integrated library system) and allow users to type in their library card number and be authenticated against the ILS database. Jim brings up a good point: if you are authenticating in this manner, you are having the patron transmit his personal information (library card number) on your WLAN. Your system should encrypt this data as it travels, so it cannot be picked out by "eavesdroppers."

What kinds of logging will you be doing (or rather, what is your equipment doing for you whether you want it or not)? Have you thought through the U.S. Patriot Act implications?

Is your network logging MAC addresses of machines that access the WLAN? Are these being erased after use, or stored? Are you having patrons pre-register their laptops' unique MAC addresses with you before they are given an encryption key or similar to log in? Are you simply having them use their library cards to authenticate to the wireless network? All of these actions can produce data, including data that can tie a specific user to a specific access time/location. Review your internal policy on user privacy! Are you violating your own privacy policy by keeping such information? If you already require that users register before using your wired Internet machines, as many libraries do, you'll want to review how you maintain that registration infor-

VPN (virtual private network)
A network that uses a public telecommunication infrastructure, such as the Internet, to provide remote offices or individual users with secure access to their organization's network. The goal of a VPN is to provide the organization with the same capabilities, but at a much lower cost. A VPN works by using the shared public infrastructure while maintaining privacy through security procedures and tunneling protocols. Data, encrypted at the sending end and decrypted at the receiving end, is sent the data through a "tunnel" that cannot be "entered" by data that is not properly encrypted. An additional level of security involves encrypting not only the data, but also the originating and receiving network addresses.

mation, and who has access to it. We would recommend using a procedure for wireless registration that is reasonably consistent with your wired access.

Will you charge for the service?

Do you currently charge for use of your Internet stations? It's best to think of wireless as simply another service you are providing, so you want to be consistent in how you provide it. If you do not charge for use of your in-house Internet stations, why would you charge for use of the wireless? If you do charge, you could allow the wireless to be free access, to encourage its use. However, that could set up an appearance of disparity or favoritism, so this is something you should discuss carefully.

Will you restrict activity by protocols or time?

Many libraries restrict their in-house Internet use by the public to http or https only, disallowing FTP, telnet, and sometimes streaming audio and video (due to the bandwidth load these take). Also to be considered are "VPN tunnels." These "virtual private networks," which provide a secure channel for transmission of private data, are created by some employers so their employees can use them on laptops when traveling to send and receive proprietary information from the employers' network. Louise has often had the experience of patrons wanting to download programs (files with an EXE extension) onto her wired Internet stations. This is blocked by her library's security system. It's a reasonable discussion to have with your IT department as to whether this restriction can (or should) be loosened for wireless access. Viruses, malware, and spyware can be uploaded to your network this way, as well, so this is a serious consideration.

Will you restrict bandwidth per user? How much?

As we will be discussing in Chapter 5, there are products for wireless LAN management that allow you to restrict the bandwidth available to individual users. If you feel the number of users you have would not overtax your bandwidth, you may choose not to worry about this, but monitor it at first and make changes based on the usage you see.

Will you be using your existing network to provide wireless Internet?

This is an excellent question, beyond the e-rate concerns (below). As discussed earlier, you may decide to:

- provide access to an entirely separate T–1 or similar for your wireless users. This may be the result of having an outside ISP providing this service, or simply to keep it physically separated from your wired LAN.
- simply have your wireless public LAN be an offshoot from

your public wired LAN, which is already physically separated (separate switches, hubs, etc.) from your staff network.
- use a **VLAN** (Virtual LAN) system to divide off your public wired and public wireless networks from your staff network.

Is your existing Internet service discounted by e-rate? Will unfiltered wireless access affect your discount?

This is a policy issue, in part based on which of the options above you choose in the previous question. Since technically you are providing Internet access in the building, even if it's not being viewed on your equipment, the general consensus is that e-rate restrictions would apply for a WLAN as they do for a LAN. There is an argument that if you're using a non-e-rate ISP for your wireless access only, you could skirt around the filtering issue, but you'd want to be sure you had sufficient disclaimers posted and a clear policy. (Please see p. 59 for further discussion.)

Some libraries choose not to filter (and thus not to take e-rate funds), so the question is, to some extent, moot. However, there are some instances where libraries choose to filter their wireless access, even if they don't filter their wired access, since they have no "line of sight" controls on what patrons are viewing (i.e., librarians can see the screens of all wired stations, providing an assumed measure of "security," but would not be able to see patron laptop screens). This is a tricky road, and we would suggest discussing this fully with policy stakeholders.

Will you provide printing to your public print stations from wireless connections? How?

You can install many network-able printers with wireless NICs to accept jobs from wireless devices. This can require creating a peer-to-peer or ad hoc wireless connection for this purpose. The majority of the libraries we viewed did not provide printing from the WLAN, but this is changing.

VLAN
A logical, not physical, group of devices defined by software, Virtual LANs allow network administrators to resegment their networks without physically rearranging the devices or network connections. VLANs are configured through software rather than hardware, so are very flexible. One of the biggest advantages of VLANs is that when a computer is physically moved to another location, it can stay on the same VLAN without any hardware reconfiguration.

The Roosevelt University library was included in the campus-wide setup of wireless access points circa 2002. . . . Going forward, students would like to be able to print to library networked laser printers using their own laptops, not currently the case. DoIT is reluctant to provide the IP addresses of these printers to the public. To resolve this problem [for now], library staff use the library's USB/Jump Drive to transfer files to a library PC for printing.
Joseph Davis
Reference/Instruction Librarian
Roosevelt University Schaumburg Campus Library
Schaumburg, IL

What our students would really like is an open wireless network so that they may connect to the Internet with their own machines. I know our technology staff is working to that end, as well as providing a way for students to access printers from their own machines. It is my understanding that there exist many network tasks in order to achieve this while maintaining a secure system. So for now, we wait.
Brian T. Johnstone
Public Services Librarian
Bucks County Community College
Newtown, PA

CREATING A TIMELINE (WHEN)

Answering the questions above (Who, What) gets you a good, basic scope for your project. It's reasonable at this point to create a basic timeline for implementation. You may be required to create such a document if you are accountable to a committee or board for your wireless project. Be sure to allow for delays in receiving equipment. If you are working with an outside vendor, make it a requirement in your contract, or in their initial proposal, that they give you a rough timeline, and update it as the project progresses. This is a necessary component of the communication issues discussed earlier. If you've told your director or board that the project will be done in six months, start to finish, you want to be sure you can meet that goal.

Authors' Note —We thought we'd share some of our own experiences:
In my library, the staff had been getting requests for wireless intermittently for about eighteen months before the library was ready to consider it, in part due to some renovations to expand the library, which would change the "shape" of any wireless layout. Once the renovations were complete, the planning stages took just a few months—and would have been much shorter, but other projects kept taking precedence. Ultimately, the money set aside for the project had to be spent, so two access points (APs) were ordered to begin with, and within a few weeks they were connected and broadcasting. The official "live" date (which happened to be tax day, April 15th) was preceded by several trial dates where the City's IT staff fine-tuned the broadcast and moved the APs to deal with interference in primary "laptop areas" that I had designated. The service continued for about

six months with just the two access points, with some areas of spotty coverage and several "dead spots" near the elevators and large concrete pillars (we have one of those big, airy, modern libraries, where huge pillars are needed to hold up the vaulting roofline). We had assumed some interference would happen, but we wanted to try it out first to see what kind of coverage we were getting. Some basic "tweaking" was done after going "live" to boost the output and adjust the lines of sight for the APs, to make them more efficient in the areas of primary coverage.

Some seven months after the initial "live" date, two additional access points were added, in particular to cover our community room area where public meetings are held. Without interruptions, the initial project would likely have taken less than six weeks, but you always have to assume that if you're using in-house staff, they will be required for other projects. If you are under a tight deadline, you might want to look at spending the money for outside contractors, who are responsible to a contracted deadline.

Louise Alcorn
Reference Technology Librarian
West Des Moines Public Library
West Des Moines, Iowa

CREATING A SITE SURVEY (WHERE)

Are you attempting to provide wireless coverage for one building or a campus-wide initiative? Do you have structural issues that may block signal, such as elevator shafts, steel beams, large cement pillars and so forth? How is your institution already wired? What cabling expansions will you need to connect access points? Will you require *wireless bridges* (large repeaters) to extend the signal from building to building or between coverage areas? The best way to answer these questions is to conduct a **site survey** to be sure you will be able to provide network access to the areas you want covered.

The Site Survey

Once you have a basic plan, i.e., where you want wireless services to be available and the scope of service, it is advisable to conduct a site survey of your library to see if and how the technology will support your plan. The site survey is a review done before installing or expanding a WLAN, with the purpose of exploring

Site survey
Review done before installing or expanding a WLAN, with the purpose of exploring line of sight issues for access points, transmission, distances for wireless transmitters, and sources of potential interference, including physical structures and other radio signals. It generally involves diagramming the network and the building, checking building plans, and testing the equipment and signal strength.

line-of-sight issues for access points, transmission distances for wireless transmitters, and sources of potential interference, including physical structures and other radio signals. It generally involves diagramming the network and the building, checking building plans, and testing signal strength in all affected areas.

The site survey can be formal or informal and can either be contracted out or done in-house, if you have an IT department willing and able or feel comfortable doing it yourself. It is important to call around to arrange for a site survey if you are planning to contract outside your organization. Many companies will only do a survey if you agree to purchase equipment and/or services from them. If you already have a vendor in mind, or one that you already contract with, they should give you a free site survey, knowing they'll be getting the resulting business.

Factors that should be included in every site survey include anything that might affect the propagation and transmission of radio signals, including elevator shafts, thick concrete walls, or interference from other nearby devices. Some building materials absorb or reflect signal. For example, materials with high water content can absorb the signal in the 2.4 MHz (802.11) range, while metal objects can reflect and cause dead spots. Site surveys should also take into account your plan for coverage, to see if a *directional antennae* on your access points might work for odd "nooks and crannies" in your building's setup. Unlike the omnidirectional antennae that an AP is generally shipped with, directional antennas can be used to target the signal more specifically, especially if the area to be covered is elongated (like down a hallway or foyer), instead of circular (like a room). If your library is located within a hospital or health facility, you may find that there are severe restrictions on the type of radio transmissions that are allowed.

Consider Channels

We've discussed the fact that 802.11b and g work in the 2.4 GHz range. However, this range is subdivided into **channels**. In the United States and Canada, this range contains eleven channels, each of which overlaps with adjacent channels. Think of going up and down a radio dial—you find that radio stations' signals can "bleed" one into the other and you find yourself hearing bits of both stations. In the same way, wireless channels can interfere with each other and lower throughput (data transmission speed) as the network tries to sort out the signals and data packets. A professional site survey will deal with these issues. In order to avoid RF interference problems, if you are using multiple access points, select channels 1, 6, and 11 (which do not interfere with

Channels
A specific band of the radio frequency spectrum used for radio transmissions.

each other because of their placement in the range). Mix them up throughout your setup so that, as best you can, you don't have two APs transmitting on the same channel to the same coverage area.

If you are in a small library and only plan to use one access point, channels should not be an issue; simply use one of the three channels 1, 6, or 11. If you are going to use just a few APs, set them to different channels spaced throughout your coverage area and you should not have a problem. For more on how to do this, we recommend Marshall Breeding's *ALA Library Technology Report,* "Wireless Networks in Libraries."

Consider Infrastructure

You must also consider where the wireless components will be connected to a wired infrastructure if that is part of the plan, including any cabling, interference or electrical issues arising from its placement in your library. According to an overview of the site survey process on the NAS Wireless Web site, a site survey should always review and determine the following:

- **The users, applications and equipment on the wireless LANs that are to be internetworked.** This details the configurations of the wired LANs already installed or that are planned for installation in the buildings or rooms that will be linked.
- **The wireless LAN system best suited for the application.** Wireless LAN (WLAN) systems are combinations of routers, bridges, hubs, and clients, as well as cables and antennas. The sight survey will determine the right combination of these components for the application.
- **The free space path requirements between antennas.** A clear RF line of sight must exist between the antenna locations. If no visible barriers exist between the LAN locations where antennas are to be erected, normally the sites can be linked. However, in some instances, a terrain analysis may be needed to ensure that the minimum required free space is available, and to determine how high the antennas must be to avoid obstructions and out-of-phase deflections.
- **The specific places where each component should be located.** Antennas must be positioned high enough for a clear RF line of sight. The wireless LAN bridges, routers, or hubs are normally placed in a computer room or wiring closet, collocated with the servers. The FCC has certified a specific cable and antenna set for each wireless manufacturer.

- **Whether to use a point-to-point or multipoint configuration.** Most Wi-Fi wireless LAN systems . . . can be configured with an option for multipoint operation.
- **Potential sources of interference in the alternative RF bands.** For complex Wi-Fi wireless LAN connections in environments where the airways tend to be busy, it may also be necessary to check for competing signals with a **Spectrum Analyzer. Spectrum analysis** can detect and measure potential sources of interference in any selected RF band.
- **The federal, state, and local regulations.** It is important to follow FCC and National Electrical Codes in the construction of masts and towers, and the electrical grounding of the Wi-Fi wireless LAN system. All designs and installations must comply with the appropriate regulations.

Questions to Ask for an Outsourced Site Survey

You may decide that trying to do such a survey yourself would be too time-intensive, or that you lack the knowledge to make a realistic survey. In that case you may look to an outside vendor. Be aware that many vendors will only do a site survey if you are planning to purchase equipment from or through them. Be clear at the outset what your expectations are for the survey. If you intend to outsource your site survey, make sure you ask the following questions prior to the visit:

- **How many individuals will be coming out to conduct the survey?** You want to know who you will be dealing with, and when they will be there. If possible, have them check in with you before and after their survey in case any questions arise that you want to be sure get answered, such as, "How will we deal with these two elevator shafts we have right in the middle of the building?" Better yet, walk them around the building when they arrive, showing them where the wired backbone for the network will be (e.g., in a server closet), where you would like to have coverage (your preferred "hot spots"), and any potential problems (cordless phones in use by staff that might use the 2.4GHz range, elevator shafts, large cement pillars, etc.).
- **Do they intend to conduct a signal strength test or just take measurements of the areas?** Whenever possible, a signal strength test is preferred. If it's an outside contractor, they should have sophisticated spectral analyzers to check for competing signals from other wireless networks, cordless phones, appliances, equipment, etc. You also want

---

Spectrum analyzer
This device, which searches a band of radio frequencies for the presence of radio signals, may be used in a site survey to determine if there are conflicting signals that might interfere with the network signal.

Spectrum analysis
This process of analyzing a portion of the radio spectrum to determine if another agency may be using it for communications can be part of a site survey.

them to check for potential "dead spots" in the building and have a good idea of the RF lines of sight for the access points.

- **How much experience do the surveyors have with wireless technologies and implementing wireless LANs?** You may have little control over this if you're using your university or city IT department, but you can mitigate their potential lack of experience by knowing what questions to ask, such as those posed in this chapter.

- **Will the surveyors provide a written plan for implementation after they have completed the survey?** Demand a written plan, preferably including a rough map of equipment placement and coverage areas. Note that this plan is not set in stone, as the actual final setup may need "tweaking" to get decent signal to all coverage areas; some AP locations may get changed slightly. This is normal. However, the surveyors should have a good idea of initial placement which they can put down on paper for you.

- **Will that plan include cost estimates and equipment quotes?** If they can give you an idea of the budget ramifications, it's well worth getting a professional site survey done. Then check those quotes against competitor's prices (usually available on their Web sites). Another important item to note: is all the equipment from a single manufacturer or vendor? Sometimes this can reduce interoperability problems, so if you find they're listing a number of **OEMs**, find out why they prefer those components, and whether they've had good luck with them working together.

OEMs
Original equipment manufacturers.

The Basics of a Site Survey

To reiterate, your site survey, whether produced by you, your IT people, or an outside contractor, should cover the following items:

- If hiring an outside vendor, **information on their company**, including insurance (bond) and experience of the main players in implementing wireless networking.
- A **timeline** for implementation.
- Any **permits required or regulations** that affect the site plan. Although this is rarely an issue, if you share space with another company or are in a clinic or hospital building, there may be some limitations and regulations to consider.
- What **standard** the network equipment will use (802.11 b/g?).

- Any **rogue wireless networks** they detected that might interfere with your network, and how they plan to deal with them.
- **Security** level/type to be used (WEP, WPA, etc.).
- A **list of equipment needed,** including access points, routers, hubs, switches, repeaters, or wireless bridges. Also, if any APs will require high-gain or directional antennae (purchased separately) rather than standard omnidirectional.
- A **list of software or firmware** you may need to purchase in addition to what will come with equipment. This would be necessary largely if you plan to have additional layers of security on your network or if they are recommending a package of wireless network management software.
- **Brand recommendations** and **price quotes** for the above.
- If they offer such a thing, get an **idea of the lifespan of the equipment.** Will you need to replace it in two years, three years? Will it need replacing due to regular wear, or because the vendor sees a new technology coming down the pike that will make it obsolete? It never hurts to ask these questions, as they help you plan for the future.
- A **map or maps,** including the following information:
  o **Tentative locations of access points** (APs). Will they be in the ceiling, on high shelves, in the open, or behind panels?
  o **RF lines of sight** – how will the APs communicate with each other through the building, and will you need repeaters in some areas to boost the signal from one AP to the next?
  o Your **channel configuration** – which APs will use which of the three primary channels: 1, 6, and 11. A general map of the channel overlap and placement is useful.
  o Based on all of the above, what the **expected signal coverage area** will be, with probable strengths (e.g., strong signal in your study rooms, but weaker signal out near the windows) and expected interference (that pesky elevator shaft). This can be simply shading on the map, to give you a rough idea of your best "hot spots." You should tell your surveyor up front where your primary locations will be, but it's possible that local interference will make access in that area difficult. If you see from their diagram an area you know will be used heavily by patrons for wireless, but which shows weak signal, you will want to ask them if there are alternate configurations that would boost signal there. If not,

you may end up choosing to move furniture to create a different "hot spot."

o **Any additional electrical or Ethernet cabling** to be run. Remember that you might decide to use POE (Power over Ethernet) adaptors instead of running both electrical and Ethernet to a location. Have the vendor cost out both options for you. And don't forget that your users will want power, too! Do you need additional electrical outlets for patron use in your hot spot?

In 2002, when the service began, laptop batteries were not as powerful as currently [2005]. Thus, students would need to use the hallway tables' electrical plugs, of which there remain precious few. Fortunately, over time, as more students bought their own laptops and their battery power increased, the electrical plug issue never became a major issue.
Joseph Davis
Reference/Instruction Librarian
Roosevelt University Schaumburg Campus Library
Schaumburg, IL

Site Survey Example: For an excellent example of a site survey report, please see **Source E** at the back of this book.

Remember that your access points will also require standard AC power, so don't forget electrical outlet placement in your site survey and mapping. If electrical power is not readily pulled to the area where you know an AP needs to be, you might consider models that use "Power-over-Ethernet" (PoE) to get their power. PoE requires additional equipment, and models that offer this option are generally more expensive. Weigh that cost against the cost of running electrical to that area.

Map Your Library

We recommend that, if at all possible, you get a professional site survey done, either by an outside vendor or your existing IT support staff. However, if you are in a very small library, or face major budget constraints, and/or you feel you have the technical expertise, you may choose to do the survey yourself. In that case, begin by creating a map of your library. Your best option is to use a copy of the actual building plans, especially with electrical markings, so you can see where power is running to plug devices into, or where large electrical convergence points might cause you interference problems. If those plans are not available, you can create your own library map on graph paper. You will then use this as a template for the site survey report. If you are using a professional site survey company, you will want to pull together copies of any building plans for them, as well, before they begin.

You may need to take some measurements before you start drawing it out to ensure that the blocks on the paper will accommodate the size of your library. Your drawing does not need to be an artistic rendering of your building; simply make sure you have a workable scale and a fairly close approximation of the building's shape. Make a separate drawing for each floor, as applicable. For example, look at Figure 3–2. In this example, the library building comprises three floors, and is mostly rectangular with a two-story addition on one side and a semi-circular lobby area on the first floor. The approximate dimensions of the building are 210 feet by 115 feet (not counting the extra length of the addition).

Next, fill in the interior spaces of your library. In the figure, you can clearly see areas marked for circulation, technical services, reference, general collection, reading room, etc. If your library has elevators or large structural elements made of dense materials (e.g., concrete pillars, etc.), make sure to include them as they tend to interfere with wireless connectivity. And don't forget to note your shelving placement—book stacks are difficult areas for uniform wireless signal. This is a potentially interfering element of a library site survey that needs to be pointed out to any outside surveyor, especially one with little experience working with libraries.

In the example in Figure 3–2, we have opted to put access points in the primary areas where patrons are likely to want to sit with laptops on the first floor. These areas include the lobby (which contains seating), the area behind the reference desk, and the reading room. On the second and third floors, only one access point for each floor is provided between the stacks and the outside wall. Inside, the stacks on the second and third floors would not be covered by wireless access.

Figure 3–2   Sample Site Map

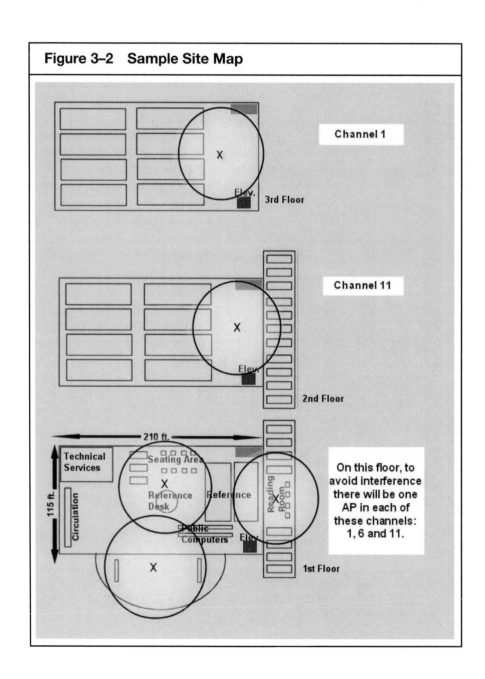

REVIEWING RANGES, BANDWIDTH REQUIREMENTS, EQUIPMENT OPTIONS, AND BUDGET (HOW)

Once your map is drawn, you can begin to get an idea of how many access points you will need based upon the scope of coverage you intend to provide. Remember, a typical access point will broadcast a radius of up to one hundred feet in any direction, with the best signal nearer the access point, depending on interfering objects. That range can be up to five hundred feet in an open, unobstructed area, but it will still be "spotty" at the edges of coverage. If you are planning for wireless access outdoors (in a plaza or between buildings in a complex), you may use access points designed for the ravages of the outdoors which can contain powerful antennas, so a greater range can be offered.

Keep in mind that wireless networks are a shared medium, so make sure you look not only at the distance that needs to be covered by each access point, but also at the area's capacity in terms of numbers of users. Use depends highly upon the activities that the users are engaging in, and the amount of bandwidth will perhaps never become an issue unless you anticipate large amounts of file sharing or users accessing high-resolution multimedia files such as streaming videos. If you decide to go with the 802.11g or the 802.11a standards, this will be less of an issue than with 802.11b. On average, you can allow for up to twelve users per access point. Areas with higher user density (e.g., the favorite "comfy chairs" where students like to study) may need additional access points to handle the load and assure each user has sufficient bandwidth. There are also software options for throttling and controlling bandwidth on your network. We will discuss these options in Chapter 4.

Obtaining a rough estimate of the bandwidth requirement for your WLAN will also help you determine the number of access points you will need for your area. Five hundred Kbps is acceptable if your patrons are only accessing the Internet. Typically, what needs to be determined is the minimum bandwidth required to provide a user running the "typical" applications at the site, with enough capacity to have a good experience. This number, multiplied by the number of simultaneous users, determines the minimum Internet bandwidth required. For example, if you determine the typical usage at your site requires two hundred Kbps of bandwidth for adequate performance and you expect no more than five users to be actively using this bandwidth at any one time (out of a potentially larger population of connected users), a one Mbps Internet connection would be required. If you are allowing streaming content, this bandwidth requirement per user would increase significantly.

Your needs assessment, site survey, and map have given you a rough idea of what you want to accomplish, what areas you want covered, and who is going to make it happen. Now you need to make decisions about what equipment you will use and how much it's going to cost. It may seem a bit late in the planning process to be reaching this point, but it really makes sense to clearly define the scope of your project before diving into exactly what components will be required and how much they will cost. Clearly, if the scope ends up extending your budget beyond where it will stretch, you will need to make revisions and cuts, but you'll be better able to make those revisions because you've already gotten a grasp of the projects priorities (staff vs. patron access, wireless labs, etc.) and where you might have some "wiggle room."

Plan the Budget

As you will see in Chapter 4, the equipment costs associated with basic wireless networking are not prohibitive to most institutions. Even libraries with just a few thousand dollars to spend can incorporate basic wireless service into their network. Earlier, we discussed other options for finding the "means" for making wireless a reality in your library, including benefactors and deals made with local vendors/ISPs. This can significantly reduce your initial costs, which are generally the bulk of money spent for such a project. Remember, however, to get everything in writing, including what the vendor will and will not pay for, support, and/or replace as needed. You don't want to save money, simply to find yourself paying all over again in a few months to have the thing professionally done, because of errors or lack of support.

You also need to keep in mind the quickly advancing nature of this technology. New standards, and devices to handle them, are coming out all the time. That is not to say you'll have to junk all your equipment in a year, but you will want to make provisions to review your equipment needs, and the available technologies, in a timely fashion (say, every eighteen months to two years). Finding equipment that is easily interoperable, especially equipment from the same or similar manufacturers, can also reduce waste, as it is more likely to survive software and firmware updates as the technology changes.

One issue not strictly in the budget, but related to it, is the "cost" of staff time. This includes your time, the time of any staff working on the project, including the IT folks, and also downtime of systems that might affect staff productivity. During your planning stage, you may want to consider a good target date for going "live" and make sure that all staff are aware that there may be issues with Internet connectivity that day. This depends entirely on how you are implementing the system, but it's something to keep in mind. If you are working with an outside vendor, be sure to ask what downtimes, if any, they foresee when you get to that point in the project. They may be vague at first, but ask the question early on, and then again closer to the proposed "live" date. They may think nothing of inconveniencing staff, but you know what a headache it can create throughout your library, so it's your responsibility to keep on top of it.

IMPLEMENTING WIRELESS POLICY

What other aspects of library service should you consider when implementing a wireless system? Two areas that may get overlooked are policy, both public policy and internal procedural issues, and the marketing of this new service.

CREATING WIRELESS POLICY – DON'T LEAVE HOME WITHOUT IT

Chances are that your library already has an Internet (or computer) use policy. If you don't, you should get one! That being said, the question becomes whether or not you need a separate policy for your wireless access. The answer to that question depends on how your existing use policy is worded. Does your policy simply give the standard disclaimer about the Internet being "an unregulated resource/medium" and patrons using it at their own risk, with perhaps some procedural information on time limits, sign-up, etc.? Does it look something like the example in Figure 3–3, from Louise's library? Most Internet use "policies" actually contain a disclaimer section, usually fairly standardized from library to library, and a procedures section, detailing that library's particular way of handling access to their Internet stations, age limits for use, printing limitations and/or costs, and so forth.

Figure 3–3   Internet Use Policy. West Des Moines (IA) Public Library.

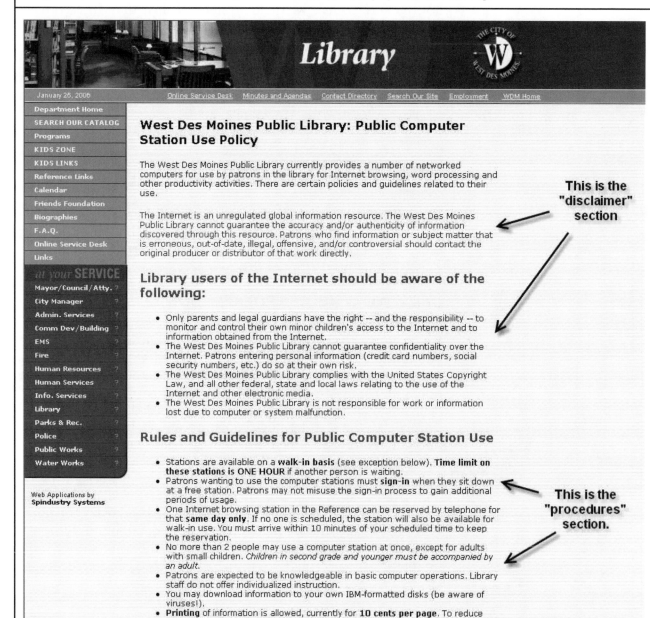

Library

THE CITY OF
W
WEST DES MOINES

January 26, 2006   Online Service Desk   Minutes and Agendas   Contact Directory   Search Our Site   Employment   WDM Home

Department Home
SEARCH OUR CATALOG
Programs
KIDS ZONE
KIDS LINKS
Reference Links
Calendar
Friends Foundation
Biographies
F.A.Q.
Online Service Desk
Links

at your SERVICE

Mayor/Council/Atty. ?
City Manager
Admin. Services
Comm Dev/Building ?
EMS
Fire
Human Resources
Human Services
Info. Services
Library
Parks & Rec.
Police
Public Works
Water Works

Web Applications by
Spindustry Systems

West Des Moines Public Library: Public Computer Station Use Policy

The West Des Moines Public Library currently provides a number of networked computers for use by patrons in the library for Internet browsing, word processing and other productivity activities. There are certain policies and guidelines related to their use.

The Internet is an unregulated global information resource. The West Des Moines Public Library cannot guarantee the accuracy and/or authenticity of information discovered through this resource. Patrons who find information or subject matter that is erroneous, out-of-date, illegal, offensive, and/or controversial should contact the original producer or distributor of that work directly.

This is the "disclaimer" section

Library users of the Internet should be aware of the following:

- Only parents and legal guardians have the right -- and the responsibility -- to monitor and control their own minor children's access to the Internet and to information obtained from the Internet.
- The West Des Moines Public Library cannot guarantee confidentiality over the Internet. Patrons entering personal information (credit card numbers, social security numbers, etc.) do so at their own risk.
- The West Des Moines Public Library complies with the United States Copyright Law, and all other federal, state and local laws relating to the use of the Internet and other electronic media.
- The West Des Moines Public Library is not responsible for work or information lost due to computer or system malfunction.

Rules and Guidelines for Public Computer Station Use

- Stations are available on a **walk-in basis** (see exception below). **Time limit on these stations is ONE HOUR** if another person is waiting.
- Patrons wanting to use the computer stations must **sign-in** when they sit down at a free station. Patrons may not misuse the sign-in process to gain additional periods of usage.
- One Internet browsing station in the Reference can be reserved by telephone for that **same day only**. If no one is scheduled, the station will also be available for walk-in use. You must arrive within 10 minutes of your scheduled time to keep the reservation.
- No more than 2 people may use a computer station at once, except for adults with small children. *Children in second grade and younger must be accompanied by an adult.*
- Patrons are expected to be knowledgeable in basic computer operations. Library staff do not offer individualized instruction.
- You may download information to your own IBM-formatted disks (be aware of viruses!).
- **Printing** of information is allowed, currently for **10 cents per page**. To reduce cost, be sure you only print out what you need.
- Regulation of the Public Computer Stations is at the discretion of the library staff.

This is the "procedures" section.

©Copyright 2006 City of West Des Moines | P. O. Box 65320 West Des Moines, Iowa 50265-0320 |Contact Directory | Privacy Policy

If your wireless access is simply an extension of your wired local area network for patron use, much of the same language from your Internet use policy will be appropriate for wireless. There are some things you may want to add, however. For example:

- "The library cannot guarantee a secure connection at all times and in all places in the library, as many things can interfere with wireless, including cordless phones, etc. The library takes no responsibility for lost data, etc. due to a lost connection."
- "Wireless networks are transmitted via radio waves, and are therefore susceptible to potential 'eavesdropping,' so patrons should be aware of what they are transmitting over this open network."
- "As the library's Internet access is filtered, so is the wireless access to the Internet" (Assuming this is the case).
- "The library does/does not provide the ability to access streaming audio or video, telnet, or FTP." These are individual policy decisions you will need to make.
- "Patrons must take responsibility for their own equipment." This is a biggie that many libraries forget. Some libraries include a legal disclaimer about not being responsible for lost items, usually copied from the library use policy, where it already exists. Laptop thefts are common, and patrons should be made aware that libraries are no exception.
- "Staff will provide no/little/some technical support, including/not including dealing with patrons' personal computer equipment."

Many public libraries offer free, open wireless access, but assume no responsibility in configuring patrons' equipment beyond simple hands-off troubleshooting (if even that). Some libraries choose to offer a minimum of assistance to patrons using their own laptops, but generally include language in their policy offsetting liability for such help. This is a tricky situation, and one you need to decide for your library and your staff. Offering no help can make library staff feel helpless, but you don't want to set up false expectations for patrons that staff would have technical knowledge of many different operating systems. See Source C in the Sourcebook for additional examples of policies, as well as sample "help sheets" which give patrons the ability to self-troubleshoot frequent problems. Help sheets are often a good solution, midway between "we don't touch patron's laptops or offer help" and expecting staff to configure patron equipment.

Louise's library, whose basic Internet policy you saw earlier (Figure 3–3), has the following addition (Figure 3–4) to their policy, relating to using the WLAN. This section is within a larger FAQ (Frequently Asked Questions) about their wireless access on their Web site at www.wdmlibrary.org.

Figure 3–4   Wireless Access Policy. West Des Moines (IA) Public Library.

Wireless Access at the West Des Moines Public Library

The WDM Public Library is pleased to provide **wireless ("Wi-Fi") access** for our patrons to our Internet service.

If you bring your own laptop computer in to the library and wish to use the Internet connection, you must have the following:

- Wireless network interface card (802.11a/b/g or compatible)
- Ability to configure laptop to use the library's connection
- Charged battery (many electrical outlets are available, but not in every seating area)
- Audio users must bring headphones. Library conduct rules still apply to laptop users.

It's useful to refer to your other, related policies with a hyperlink.

Don't forget to mention theft liability.

Policy and Procedures:

Library staff members will not assist customers with their computer or their configuration. Library staff cannot accept the liability of handling non-library equipment.

The library does not assume responsibility for any damage, theft, alterations, interference or loss of any kind to a user's equipment, software, data files or other personal property brought into or used at the library's facilities.

All virus and security protection is the responsibility of the user. Our Internet connection is open and unfiltered. Be aware that as wireless access is sent over radio signals, 'eavesdropping' by other users, though unlikely, is possible. Keep this in mind as you choose whether to transmit confidential data.

The library cannot guarantee that this service will be available at any specific time or at any specific speed, nor can the library accept reservations for wireless access.

The connection shall not be used for illegal or time- or bandwidth-consuming commercial purposes. Users are asked to sign off the network when they're done, so that IP address can be freed up for other patrons.

Printing is not available through the wireless access. Users who want to print will need to save to a diskette or removable (USB) drive and use a public workstation to print.

All other applicable rules from the WDMPL Internet Use Policy will apply.

Additions to computer use policy relating to Wireless Access.

How do I use wireless at the library?

You will need to bring your own laptop computer with a WIFI network card to the library. The library access points transmit on the Wi-Fi standard IEEE 802.11a/b/g. Any standard device capable of accessing on this standard should be able to connect.

Will I need any special settings or passwords to connect?

The library's wireless Internet access is open to all visitors. No password is required. If your laptop is **configured to obtain an IP address automatically (via DHCP)** you should be able to connect immediately. If your computer does not connect, use the suggestions below under Wireless Card Settings. Because each person's laptop is different you will be responsible for knowing how to configure your own equipment.

Where can I connect my laptop?

The Wi-Fi signal will work most places in the building, depending on interference. Some areas are better than others. Feel free to ask staff about strong signal locations. **Electrical outlets** are readily available throughout the building.

In this case, the policy is mentioned as part of a larger FAQ on using wireless at the library.

If your wireless network is completely separate from your staff network, you have saved yourself some security issues. However, your users still need to be told what risks they take using the service, what the acceptable use of this service is, and what it means to them to use it with their own equipment. In an article for *Centralities*, the Central Massachusetts Regional Library System's newsletter, in February 2005, technical specialist Rick Levine discussed such a situation (specifically a library that had received free wireless from a local broadband ISP) and argued the following:

> . . . since your cable connection in no way "touches" the library's network, opening it to the public involves little more than attaching a consumer-grade wireless router. However, even while this arrangement leaves your library's computers protected, it does expose wireless users to all the security concerns engendered by wireless networking. It is therefore important to develop a policy that delineates these concerns and advises patrons of terms of use.

Some of the main points Levine suggests should be addressed in a wireless use policy are:

- **The wireless connection is not secure.** Due to possible interception, patrons should be careful about transmitting credit card numbers, passwords, and other personal data.
- **The wireless network is open,** and connected computers can be vulnerable to other users' viruses or malware. Users are strongly encouraged to update their antivirus software, firewalls, etc. before accessing the wireless network.
- **Library staff cannot offer technical support** for your laptop. The library is not responsible for damage to hardware or software, loss of data, or theft of unattended equipment. Keep your laptop with you at all times.
- **Any limitations of your WLAN,** such as Web-Based e-mail only, no printing allowed, no streaming audio/video, etc. should be mentioned. Some routers allow you to configure access options (e.g., maximum simultaneous users, time of day restrictions), which should also be mentioned in any policy/FAQ. One item which Levine does not mention, but which you should if this applies at your library, is if patrons cannot use Outlook and similar software to pull their e-mail due to no SMTP server being attached, etc. This is a common limitation, but often patrons don't know about it until they come in and have trouble.

- **All applicable library policies apply,** including legal and acceptable use of computers and the Internet. If you are posting this FAQ on the Web, you can provide links to appropriate policies.

Figure 3–5 Wireless Policy Disclaimer. St. Joseph County (IN) Public Library.

Limitations and Disclaimers

The Library's wireless network is not secure. Information sent from or to your laptop can be captured by anyone else with a wireless device and the appropriate software, within three hundred feet. Library staff is not able to provide technical assistance and no guarantee can be provided that you will be able to make a wireless connection. The Library assumes no responsibility for the safety of equipment or for laptop configurations, security, or data files resulting from connection to the Library's network. Printing capability is not available on the SJCPL wireless network.

Ultimately, you need to be sure that your wireless access policy, whether a separate policy or an extension of your existing Internet use policy, reflects the mission of your library, as well as your existing policies not only on Internet use, but also appropriate library conduct and use policies. If you are offering assistance, or going so far as to check out laptops for patrons to use in the building (as many libraries are beginning to do), be sure that your policies and procedures are in place before you go "live" with wireless. You don't want what should be an easy, useful new service for your users to become a source of annoyance or additional problems for staff. Policy should be reviewed by your board or advising committee, and preferably your university or city attorney, to be sure the language is appropriate both for liability and is also in line with your other existing policies.

Don't forget to let your staff know what the policy is. This sounds silly, but wireless initiatives can happen so quickly that staff may not have time to become aware of all the issues involved, especially what patrons will ask them. Figure 3–6 is an excellent example of a pre-launch "policy" document, which is really a combination policy and FAQ.

Figure 3–6   Wi-Fi Policy. Howard County (MD) Library.

Close this window.

Howard County Library WiFi Policy
Author: Amy Begg DeGroff
Publisher: Howard County Library
Copyright: 2005 Howard County Library
Date Posted: Jul 25, 2005

This is an <u>excellent</u> FAQ for staff, including both policy and informational items presented in a non-threatening format.
Image was edited for space. Full text available here:
www.webjunction.org/do/DisplayContent?id=11059

Description: A friendly and informative handout helping staff support wireless patron computing, from a suburban Maryland library.
Fellow HCL employees

Hello and happy summer!

I see and hear that every day, more and more customers are taking advantage of our Wi-Fi services. The result is that many of you are fielding questions about access which you may not be comfortable answering.

Below are a few points to share with you about the Wi-Fi configuration. This information will be posted to the Staff Intranet in the next few days.

(1) **The extent of troubleshooting HCL employees can provide for the Wi-Fi is to confirm that it is available.**   This confirmation is accomplished by using a Wi-Fi enabled device to launch a web browser and reach a web site. In order to do this, we will deliver to each branch, a low level laptop with Wi-Fi capabilities. HCL employees can pull out this laptop and confirm the ability to reach the Internet via the Wi-Fi.   ──→ **This is an excellent idea. Make sure that staff get trained on its use.** ←──

(2) IT staff will not provide technical support on customer computers. No exceptions. **Do not page or call IT for Wi-Fi support.** We cannot troubleshoot computers that are not part of our inventory.

(3) Finally, HCL employees are not permitted to configure or troubleshoot a customer computer. No exceptions. **Do not touch a customer computer.**

A few more detailed points....

Why does HCL have Wi-Fi ?
HCL decided to offer free access to Wi-Fi as a way to provide a public service to the community. We recognize that many of our customers own computers (and many are laptops) and want their information and research all in one place. Also, we did moved to Wi-Fi for selfish reasons -- anyone bringing in a pc to work is not using one of our pcs -- freeing up our pcs for other customers.

How is the Wi-Fi configured?
We configured our Wi-Fi to provide access to the Web and to provide access to information. The Wi-Fiis not as open as our internal network nor is it as open as someone's Internet access at home or at their place of employment. It might also not be as open as another free Wi-Fi provider.   ──→ **An excellent point for staff to know.**

How does someone use the Wi-Fi?
The walk in and their device (laptop or PDA) finds the network and VOILA-- they are good to go.

What is the deal with EMAIL via the HCL Wi-Fi?
We've shut down SMTP in order to protect our infrastructure and our customers from a SPAM engine being brought in.

... Amy, you now sound like an adult in a *Peanuts* tv special.
If the email is set up to move messages from a server to a client based email application (such as Outlook) it will only partially work on the HCL Wi-Fi.   ──→ **This is common with library wi-fi. However, many patrons assume that since they're using their own laptop, their access will not be restricted in any way. Make it clear!**

... Amy, what the heck does that mean?
Email can be RECEIVED AND READ either via a web browser (like Firefox or a Internet Explorer) or via a client - such as Thunderbird, Eudora, Outlook..) If a Wi-Fi user is trying to SEND email from a client it will not work. Wi-Fi users MUST send email via a WEB BASED email client.

... Goodness, Amy, how will I know?
The way I figure this out is to say " are you using Firefox or Internet Explorer to send email?" if they say, "no, I am using Outlook Express" you have your answer.

... Amy, more detail, please?
When you read email via OpenWebMail you are on a web browser. If you use Thunderbird at your desk, that is an email client. Is this getting more clear?

... Hey, I am almost getting this...
If you have ANY software on your computer (other than a web browser) to read email, you are using an email client and you WILL NOT be able to send email via the HCL Wi-Fi. You will be able to read it and receive it. You can also write and email and put it in a DRAFT folder to send later.

If you are ONLY in a browser, your email should work.

What if there is another problem? What if they can't find the Wi-Fi on their computer?
You can always suggest the tried and true step - reboot. I find with my laptop, when I turn it off and then back on, it finds the Wi-Fi network. Otherwise, the customer needs to contact their computer support team.

Communicating Policy to Users

Any policy you create is designed not only to protect the library, but also to inform your users. Therefore, you want to make an effort to communicate this policy to users. One option for this is simply to post the policy in your "hot spots" as well as on your Web site. You can also have patrons manually register with your staff before use, and require them to read the policy before issuing them an encryption key or password. A combination of these methods is also possible.

A technological option for transmitting policy text to users is the captive portal. This is a method of "capturing" your Web traffic and forcing users to "click-through" a user agreement before they can access the Internet at large. If you've ever used the Internet service at a Starbucks or a Borders store, you have encountered a captive portal. Basically, it forces an HTTP client on a network to see a special Web page before being released out to the Internet. This is done by intercepting all HTTP traffic, regardless of address, until the user is allowed to exit the portal. Usually captive portals are used for authentication. At the commercial "hot spots" like Borders, which along with Starbucks uses the "T-Mobile Hotspot ™" ISP, you have to purchase a day pass or subscription and register with their service in order to proceed. Without agreeing to, and paying for, their service, you can go no further on their wireless broadband access.

Libraries use captive portals not necessarily to procure payment, but to limit access to users who match a criterion—those who are pre-registered, those who are library card holders, students or faculty, or simply those who accept a user agreement. This last limitation is particularly useful for libraries. They can require that before a patron accesses the WLAN for the first time, he or she must read and, by checking a box or similar, accept a user agreement of the library's design. The library can also, at this point, require that the user register and/or authenticate against a database (e.g., its ILS or a database of student and faculty ID numbers) , but you don't have to. According to Wikipedia.org:

> Captive portals are gaining increasing use on free open wireless networks where instead of authenticating users, they often display a message from the provider along with the terms of use. Although the legal standing is still unclear (especially in the U.S.A.) common thinking is that by forcing users to click through a page that displays terms of use and explicitly releasing the provider from any liability, any potential problems are mitigated.

As you can see, a captive portal serves as a means of authentication as well as an informational tool to supply a user with policy text. Captive portals can be either hardware or software solutions, and can also be used on wired networks to secure, for instance, open Ethernet jacks. Many WLAN hardware providers (Aironet, Cisco) have captive portal or similar solutions, some of which are listed in Source B. It's worth discussing with your hardware vendor which products can easily include a captive portal function.

As mentioned elsewhere, be sure you are aware of what patron information the portal is transmitting, and whether that data is secure; and also what information the portal is saving, and whether the storing of the data is in line with your patron privacy policies.

Figure 3–7   Checklist for Wireless Access Policy.

First, look at your existing Internet (or Computer) Use Policy. Do you need to add anything to it relating to use of the wireless? You may decide that it covers your situation. Do keep in mind the following possible additions, however:

- **Network Security**: If you're providing a fairly open network, include a disclaimer about the possibility of radio signals (wireless) being intercepted. This is more specific to wireless than the disclaimers in your Internet policy about "library is not responsible for lost data due to network failure" and "beware of viruses" and "be careful about transmitting your personal information on an open network".
- **Network Availability**: WLANs can be flaky, patron laptops only more so. Note that they may lose signal at random, and the library takes no responsibility for lost data, etc.
- **Limitations on Use**: Time limits, bandwidth limits, no FTP, no telnet, no streaming content. Do you offer printing? Web-based email only (no SMTP server)?
- **Personal Equipment Security**: Warn patrons that the library is not responsible for stolen equipment, lost data due to their equipment failure, etc.
- **Filtering**: Note if the wireless access is filtered, especially if the in-house is not, or is only partially filtered (filter by patron choice only, for instance). You may want to briefly quote any law (CIPA) relating to this.
- **Support**: Will your library staff provide help with patron laptops? Can they provide help with determining if there is a signal present (i.e., if the APs are working)? If you don't want staff touching patron laptops due to liability, say so.
- **Staff**: Make sure your staff are kept "in the loop" about any wireless initiatives, in particular about what they'll be expected to offer in the way of support for patrons.

We are a library in a town of ten thousand, faced with the usual problems this represents. We have a wide span of patrons that want just about everything –including technology –but [we] try to live on a limited budget. Wireless solves some space and money issues.

We've been wireless for about a year. We don't really have any problems. Truck drivers pull up in the parking lot and use the Wi-Fi. Some don't even come inside. Students and patrons bring in their computers and use them wherever they land. No issues really. We're happy.

We are freely available. Anyone that comes in must abide by our Internet policy –so we don't even have a special policy for wireless. So far we just haven't needed one.

We've been queried by several libraries considering going wireless. They all have questions and concerns and fears. I just tell them how we're set up and that it isn't that big a deal. It does scare a lot of people though. Our administration is just fearless!
P.J. Capps
Director
Atchison Library
Atchison, Kansas

MARKETING YOUR WIRELESS SERVICE

"Where can I use my laptop to access the Net?"

"Do you have wireless? I need to upload some data from my laptop to my employer."

"When are you guys going to get wireless here, like they have at Borders and Panera?"

You probably hear some variation on these questions several times a month, especially at a public library. More and more, patrons are coming to expect certain basic services from libraries, now including free wireless Internet access. At academic libraries many students arrive on campus with wireless-capable laptops in hand, expecting to be able to check class schedules, upload assignments, and e-mail home, all from the "comfy chairs" in the back corner of your library. At public libraries, patrons may arrive from out of town, wanting to use their laptops to do work somewhere other than their hotel room, or to download their latest distance-learning class assignment. Or just check their e-mail from the "comfy chairs" in the back corner of your library. See a pattern emerging?

In response to these requests, and others, you've decided to implement a wireless project. Great! Have you thought about how you will let your library patrons know about the service? "But they'll know next time they come in and try to use their laptop!" you say. Well, yes, some of them. But what about new users? What about that guy who comes to your library every day, but assumes you don't have wireless, so leaves his laptop at home? What about your library board, city council, or faculty oversight committee? They may be aware, thanks to budget requests, that you are implementing wireless, but do they know how to use it? Do they know when it is scheduled to go "live?"

Marketing may be the least of your concerns when you're trying to get wireless service "off the ground," but you should not forget it entirely. Have you considered "partnering" with local groups, including your library's Friends group, to promote the service? If it's a campus-wide initiative, have you called the university newspaper to have one of their student reporters come over and "cover" the new service? If you're in a public library, has your local newspaper done a "where are the Wi-Fi hot spots" piece? Were you in it? You send out press releases about adult and children's programs you want included in the newspaper— why not a new service? Figure 3–8 shows a great, positive article about the Tonganoxie (KS) Public Library's new wireless service. This kind of publicity is the kind of thing that makes patrons (and library boards!) take notice.

Figure 3–8 Wireless Service. Tonganoxie (KS) Public Library.

www.tonganoxiemirror.com/section/archive/story/8294

THE MIRROR
Internet Edition

Reader Forum   Area classifieds   Weather   Add your organization   Search

The press can be helpful with your library marketing!

sections
classifieds
top stories
sports
viewpoints
living
education
remember when
obituaries
photo galleries
movie reviews
almanac
election 2005
government
public notices

send us your news
feedback
news tip
achievement announcement
calendar event
birth announcement
anniversary announcement
engagement announcement
wedding announcement
obituary
letter to the editor
press release

community
groups & clubs
area links
religion
government info

assistance
subscriptions
search our site
archives
contact
about the Mirror
advertising

News neighbors
Baldwin City
Basehor

Archived stories

✉ E-mail story   🖨 Printer Friendly   ✉ E-mail editor

Library online as city's first hot spot
By Shawn Linenberger, Reporter
Wednesday, September 28, 2005

Bring your laptop and prepare to surf at the Tonganoxie Public Library.

Last Thursday, the library became Sunflower Broadband's newest hot spot. The library, which is at Third and Bury, now has wireless, high-speed Internet access. Residents can use their wireless-capable laptops at the library and surf the Internet.

"We think it's good for business people, students, visitors to town who bring their laptop and need a place to connect," library director Sharon Moreland said.

advertisement

THE MIRROR

Get the Mirror at a great rate

All your local news, photos and information - for as little as $30 a year!

Order online at our secure Mirror store!

Moreland inquired about the library becoming a "hot spot" after she attended the Northeast Kansas Library System Technology Day last month in Topeka.

"We were discussing social technologies and it was mentioned that the Eudora and Lawrence libraries were hot spots," Moreland said.

The next day, Moreland contacted Sunflower Broadband about making the library Tonganoxie's first hot spot.

The library's only cost for the hot spot was for router configuration, which cost $187.

"It allows more public access computers at no cost to the taxpayers," Moreland said.

Sunflower donation   Make sure your benefactor(s) get mentioned!

Ashley Seeger, who works in customer support at Sunflower Broadband, said the company provides the hot spots to local libraries as a community service. Internet service and equipment is donated to the libraries, Seeger said.

From the perspective of the librarian who coordinated publicity, FAQ sheets, demo-ing it to staff, etc. it was very easy and mostly trouble-free. . . . I've seen as many as eight people at a time here with laptops. The local papers each gave us a blurb about the service, and the City of Madison put info about it on their press release page.

Liz Amundson
Reference
Madison Public Library
Madison, WI
[www.madisonpubliclibrary.org/services/wireless.html]

Figure 3–9   Map of Electrical Outlets. Madison (WI) Public Library.

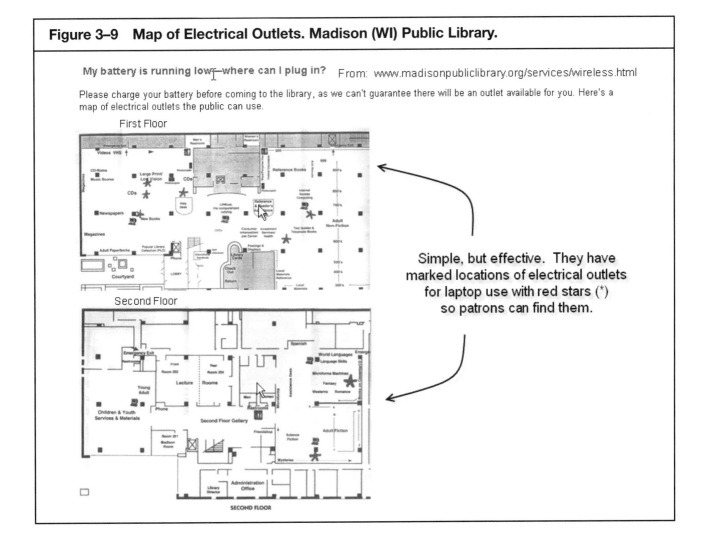

My battery is running low—where can I plug in?   From: www.madisonpubliclibrary.org/services/wireless.html

Please charge your battery before coming to the library, as we can't guarantee there will be an outlet available for you. Here's a map of electrical outlets the public can use.

Simple, but effective. They have marked locations of electrical outlets for laptop use with red stars (*) so patrons can find them.

From a customer service perspective, what information are you giving out to patrons? Are you simply announcing that you have wireless and hope patrons "figure it out" (many of us have done this), or are you being more proactive? The Madison (WI) Public Library has an extensive FAQ on its Web site about its wireless service. One innovative addition is a *map* of where electrical outlets are located in their building for patrons to plug in their laptops (see Figure 3–9)! In a public service environment like a library, good customer service is your best marketing tool. Be proactive and responsive. See Source C for more examples.

Figure 3–10   Checklist For Marketing Your Wireless Service

- **Tell your staff first**. Weeks or months in advance of WLAN installation, be sure all library staff are at least roughly aware of the imminent arrival of this new service. Give them a basic explanation of what it is, what patron needs it will improve or expand, and what their role in it will be. Will they be required to reset access points? If so, note when they will be getting training on this task. Allay their fears and you'll have much more support from staff as the project moves along.
- **Tell your larger organization**. Are you a university library, a public library that is part of a school or city structure, a corporate or medical library in a hospital or other organization? If so, be sure you let the larger organization know what you are planning. This may be part of the budgeting and approval process, but it's a good idea to let fellow employees know about this new service once it's imminent. Obviously this is dependent on the sort of relationship you have with the larger organization. If you tend to come to virtual blows over every perceived difference in service levels, then you might leave it be. You'll have to decide for yourself. However, if you have a reasonably healthy relationship with faculty, staff, city government and the like, let them know what's coming and let them help you spread the good news to citizens, students and other potential patrons.
- **Tell your patrons**. You'd be amazed how many libraries forget to let their patrons know when their wireless access goes "live." In many ways it feels like the end of a project, but really it's the beginning of a new service. As discussed above, you have many options for letting them know: post it on your Web site as a news item, link to a FAQ with all the pertinent details (more on this in Chapter 5 and Source C, put up tent cards in areas you've designated as "hotspots", and create handouts for patrons and staff.
- **Make a map**! A map of your "hotspots", and also where your electrical outlets are located. This map is helpful to both patrons looking for service, and staff trying to explain to patrons where they can plug in their laptop and get online. If you require them to register and get an encryption key (see page XXX), this is a great opportunity to hand them a copy of your Wireless FAQ and a Hot spot Map!
- **Tell your colleagues in the greater library world**. Do you have a state-wide library e-mail list? A regional or local e-mail list comprised of librarians? Send a small announcement via e-mail to let them know about your new service. This can have two effects: they may send patrons your way if they do not have the service or if it's down, and they may ask you how you did it because they want to provide such a service themselves. Share the wealth of experience!
- **Tell the press**. You send out press releases about upcoming programs, special events and guest speakers. So send a press release about this new service you're providing. If you're a bit wary about how it's going to work, wait until it's up and running stably for a few days, then fax your local newspaper, student newspaper, and TV stations. It's possible they're working on a "Wi-Fi hot spot" piece, and you can get in on the free PR, as Thomas Edelblute did.

During the second-floor remodeling of the Central Library project, we installed a public wireless hot spot in our Quiet Zone. It is a simple wireless router connected to a separate DSL line so I would not have to worry about the security issues on our network. There are VLANs that can secure our network, but that would require a more expensive network infrastructure than I can pay for at this time. To extend the range, there is a second access point further back in the quiet zone. We opened this up in 2004.

The public found it before we advertised it and has been a hot item ever since. Then one night the information desk received a phone call asking if we provided wireless access. The librarian on desk read off information from the flyer we have. When asked if there was any more information they wanted, the voice on the other end said they were with the local newspaper and doing a story on wireless.

"Do you want to talk with the City Librarian, who can provide you with more information?" the librarian asked.

"No, I have all the information I need."

This scared the librarian because we had no idea what to expect in the newspaper and alerted the City Librarian, the City's Chief Information Officer, and myself.

The next day the *Orange County Register* had an article about wireless hot spots around the county, with a sidebar listing the locations of various hot spots in the county. We are first on the list.

Thomas Edelblute
Public Access Systems Coordinator
Anaheim Public Library, CA

See Chapter 5 and Source C for a longer discussion of what should be in your FAQ and Troubleshooting documents for patrons. Also, in The Wireless Sourcebook at the end of this book, you will find links to and samples of such documents from a variety of libraries in Source C.

4 HARDWARE EQUIPMENT AND INSTALLATION

TEN BASIC QUESTIONS TO ASK BEFORE CHOOSING EQUIPMENT

The kind of equipment you decide to purchase will depend on a number of factors we've already touched on, including the service you plan to offer, scope of the installation, and the primary purpose of the network. Many of these factors will be addressed in your site survey, with equipment options being presented as part of the final output. This chapter will revisit some of these factors and how they affect hardware choices. In particular, we will pose the following questions and discuss how they can be answered with hardware (and software) solutions:

- What wireless network (WLAN) topology do you use? This is your WLAN configuration of access points, any switches or repeaters, the wired backbone and peripherals.
- Are you creating a network or networks for both staff and public use? How will these be separated—physically or using VLANs (virtual LAN segmentation).
- How will you address roaming issues? Will staff need to

roam between the staff and public side WLANs? There are some proprietary solutions to roaming, but you may need to buy same vendor equipment.

- Are you planning to provide printing? Do you have a printer with a wireless NIC? If you charge for printing, can you run the print management software with this printer, too?
- Will your staff use require the use of wireless peripherals like barcode scanners to do inventory?
- Are you planning on creating a mobile lab, or planning to lend out laptops for patrons to use? What equipment specifications will these devices need?
- How will you achieve interoperability of all equipment you plan to install? Do you have an existing contract with a vendor? Will this limit your equipment choices?
- How can you find certified Wi-Fi equipment that you can trust to be interoperable?
- Do you want to limit the range of your wireless network to certain areas, certain bandwidth? What WLAN management tools are you considering?
- How will you handle installation? Take into account the placement of equipment (in ceilings, etc.). Do you need network cabling or electrical run to these areas? Can you use Power Over Ethernet (PoE) adapters to save money?

CLARIFYING WLAN TOPOLOGIES: BASIC CONFIGURATIONS

Although we covered some basics of WLAN configurations in Chapter 1, we want to revisit them here, to clarify where hardware fits into your average WLAN setup. The 802.11 standard allows for a few basic configurations, or "topologies," of equipment.

IBSS: INDEPENDENT BASIC SERVICE SET

The simplest configuration is called an **independent basic service set (IBSS)**. In this configuration, only the endpoints or stations communicate with each other in a peer-to-peer-style network, often referred to as an **ad hoc topology**. In a peer-to-peer or ad hoc network, there is no server, access point, or centralized workstation that coordinates and controls communication and access. The dataflow in this case goes directly from one station to an-

Ad hoc topology
Also known as **independent basic service set (IBSS)**, this is a wireless network framework in which devices can communicate directly with one another without using an access point (AP) or a connection to a regular network. Different from the more common infrastructure network in which all devices communicate through an AP, libraries would rarely use this mode, except in some staff and inventory projects, or for connecting wireless peripherals as needed.

other without any kind of repeating or signal enhancing device in between (see Figure 4–1). Peer-to-peer networks are, by nature, insecure and not recommended (or even practical) for most applications. For the most part, these types of arrangements are quite temporary and allow two individuals to share data between their wireless devices. You may have seen two PDA users "beaming" data to each other. This is the same idea.

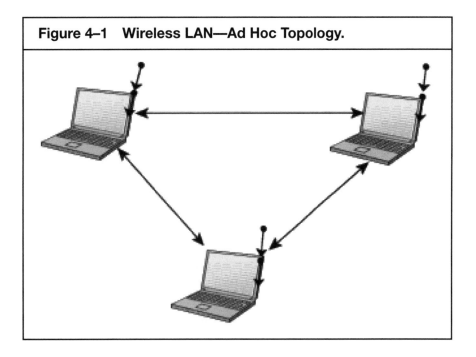

Figure 4–1   Wireless LAN—Ad Hoc Topology.

Libraries might use this configuration for transfer of data from inventory devices, but would not, as a rule, use this for wireless access for the public.

BSS: INFRASTRUCTURE BASIC SERVICE SET:

The second type of WLAN configuration, known as infrastructure mode, or an **infrastructure basic service set** (BSS), includes an access point that relays data from one station to another and functions as a buffer and possible gateway in between. Access points in this type of environment serve to expand the coverage area because the stations don't have to be as close to each other as in ad hoc mode. With the access point in a central location, each station can be up to one hundred feet from the AP, or as much as two hundred feet from the other stations while still main-

Infrastructure basic service set (BSS)
A type of IEEE 802.11 network comprised of both stations and access points (APs) that are used for all communication within the BSS, even if the stations reside within the same area. Communication from station to station takes place through the APs.

taining a central connection. More access points can be added to cover larger spaces. More importantly, the access point's ability to function as a gateway allows the WLAN to interface with a wired network.

ESS: EXTENDED SERVICE SET

If your particular configuration will require the installation of multiple access points, you will then need to decide if those separate access points will each control their own BSS, or if you will enable users to roam from one access point to another in a reasonably seamless fashion. In the first instance, users are confined to the area around their immediate access point. Once they stray out of the coverage area, they will lose their network connection even if they come within range of another access point. In the second case, users can move throughout the entire WLAN area, from one access point to another, without losing their connections at all. This latter configuration is known as an extended service set (ESS), and is a likely choice for most libraries using more than one AP. In an ESS infrastructure, each access point acts as a sort of relay station, determining if data should be sent to stations within its BSS, or passed on to other access points or perhaps to the wired network. (See Figure 4–2) In their book *802.11 Demystified*, James and Ruth LaRocca describe this service as "the mechanism in IEEE 802.11 . . . called the distribution system (DS). The DS is a logical concept. It is independent of the media connecting an AP to another element of the DS. This enables an AP to communicate to the stations via radio while having a choice of media (radio, wired LAN, and so on) for communication between APs or external networks." (LaRocca 2002, 25) To the user, the entire framework (including wired and wireless LAN) appears as a single unified network. Chances are good that this is the configuration you will want if your wireless network uses more than one access point and is tied to a wired backbone. Added to this may be a connection to an authentication server of some variety, against which potential users are verified when connecting.

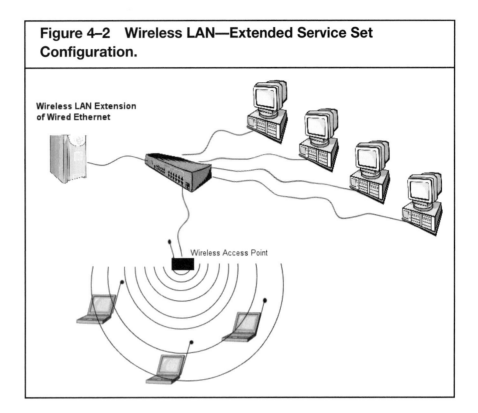

Figure 4–2 Wireless LAN—Extended Service Set Configuration.

Wireless LAN Extension
of Wired Ethernet

Wireless Access Point

CREATING WIRELESS STAFF AND PATRON NETWORKS

In most cases, the layout of your wireless network is related to the layout of your wired network. The first question you must ask is whether your staff and patron wired networks are currently separated. Separation of these networks protects your library's data infrastructure from dangerous insecurities. We will discuss the security implications more fully on page(s) 100, but for now we will discuss how your wired layout will affect your wireless configuration. Your staff and public-side networks can be actually physically separated, using separate switches or hub ports, or may simply use a VLAN (Virtual Local Area Network) arrangement so the staff network's data is not viewable to patron network users. You may also have chosen (as many libraries have) to put your patron network on the outside of one or more firewalls, or at least in the "DMZ" (demilitarized zone—a semi-controlled interim "space"), where you assume some amount of deliberate or accidental mischief is possible and where it can be controlled. Your patron data, which is stored on your ILS and/or elsewhere in your staff data files, can thus be segmented off behind a layer of protection. This is important, as this data may be legally protected by state laws or regulations on privacy.

When considering adding wireless access to your library, keeping staff and patron networks as separated as possible makes the implementation of wireless extensions to the network that much easier. If you are planning on wireless for the public only, you can simply make the wireless an extension of the public network, so that providing open access (as you may want to do) will not endanger your staff network. The security that protects your staff network from mischief on your patron network would also protect it from the WLAN extension to that patron network. In the same way, if you decide you need a staff-side WLAN, you can decide whether they will simply use the public WLAN—reasonable if you simply want them to have Internet access at some remote location—or if you will have a secure WLAN that provides access to information on the staff network.

How does this affect hardware considerations? Let's say that you just want to provide wireless Internet access for patrons coming in with their own wireless devices. If you have a separate public network from which to extend your wireless access, especially if it has a full physical disconnect from your staff-side network, you can effectively plug in an access point (or points). Depending on the level of security you want to set for that WLAN, you can be off and running fairly quickly. That is, obviously, quite simplified (though in fact a WLAN at home is about that easy) and you will have to consider placement of the APs, securing of the APs (on the walls or ceiling, behind a door), electrical run to the APs, etc., but you get the idea. However, other staff/public access needs may require more creative thinking when considering wireless.

If you have need for both public and staff to access the Internet wirelessly, as stated before, you could simply make the connection from the public side and have the staff use this connection for simple Internet access. However, if you need staff to be able to access material on your secure network while out working with the public, this gets trickier. Your best option is to use the VLAN approach, which you may already be using on your wired network. You can create a virtual LAN and limit it to the MAC addresses of staff devices that will be connecting to it, or a particular IP subnet. You can require an authentication process so that staff have to sign on to the staff-side WLAN as they do to the staff-side wired network. If this situation applies in your library, then these are issues you will want to discuss at some length with your network or system administrator, or your technical support person/crew. One major topic of conversation should be whether you will have staff roaming from public to staff-side and back again, which adds a whole extra level of complexity and security. We will discuss some roaming issues below, but as a caveat, something as complex as the situation we just described really requires a high level of network expertise to implement securely.

ADDRESSING THE ROAMING ISSUE: MOVING BETWEEN ACCESS POINTS AND SUBNETS

In many wireless environments, users will not remain static in their locations, but will want to roam while maintaining a continuous connection, perhaps moving from one access point to another. This is especially important in a facility where staff are using wireless connections to provide point-of-service help (reference, inventory, circulation, etc.). Most wireless LAN equipment supports this basic kind of roaming through a process by which the end-user client automatically associates with the new access point. Robert Keenan observes in an article for CommsDesign.com:

> Roaming in WLAN networks has been anything but easy, especially when real-time applications like voice are involved. Typically, every time a Wi-Fi user roams from one AP to another, they must re-associate, go through another authentication process, and perform a key exchange — a long process that can cause data and real-time connections to be lost (Keenan 2004).

He describes proactive key catching (PKC), a solution offered by the company Airespace as "a proprietary extension to the 802.11i security spec . . . that allows users to more freely roam in a WLAN network." He then explains how it works: Through PKC, users are issued a key during the initial authentication process. As the user roams, the access point, which links up through a WLAN switch to a Radius server, identifies the key, and if it recognizes the key can allow the user to continue a connection without going through a full authentication process (Keenan 2004).

Although this roaming problem is less of an issue within a single

Subnet
A portion of a network that shares a common address component. On TCP/IP networks, subnets are defined as all devices whose IP addresses have the same prefix. For example, all devices with IP addresses that start with 100.100.100. would be part of the same subnet. Dividing a network into subnets is useful for both security and performance reasons. IP networks are divided using a subnet mask.

subnet, if your wired infrastructure, and by extension your WLAN, is configured with multiple subnets, this can create roaming problems. Jim Geier describes one manufacturer-offered solution to this issue:

> With multiple subnets . . . mobile users must be able to seamlessly roam from one subnet to another while traversing a facility. WLAN access points do a great job of supporting roaming at Layer 2. Users automatically associate and reassociate with different access points as they move through a facility. As users roam across subnets, though, there must be a mechanism at Layer 3 to ensure that the user device configured with a specific IP address can continue communications with applications . . .
>
> Mobile IP, offered by some access point vendors, solves this problem by allowing the mobile user to use two IP addresses. One address, the home address, is static. The second address, the "care-of" address, changes at each new point of network attachment and can be thought of as the mobile user's position-specific address. (Geier 2003)

What you will need to consider in your network planning is how your wired infrastructure is divided into subnets. If two access points are on separate subnets, traffic will have to cross a router, something that most wireless LAN vendors currently do not support. This is where good pre-planning is useful (see Chapter 3), as you can "map" out how your wireless access will be used, and how the wireless access will be affected by the existing wired infrastructure, before you begin installing hardware.

PROVIDING PRINTING ACCESS

Are you planning to provide networked printing access from the wireless network? If so, you will need a printer or print server with a wireless NIC (Network Interface Card) that can handle the sends. Very few of the libraries we spoke with had implemented printing options for their wireless networks. Peripherals like printers with Wi-Fi capability are still in early stages. However, if you want to provide this service, there are a number of options currently available. A wireless print server is a small computer and Wi-Fi radio built into a single box, much like an access point or gateway, that can enable your printer to access your network. A Wi-Fi-equipped printer connects directly to your Wi-Fi network.

There are several printers on the market now that are specifically designed with an internal NIC and the capability to interface with Wi-Fi networks to handle print jobs. A search at Wi-Fi Alliance's excellent product referral site (discussed further in Chapter 3) produces several dozen models from a number of name-brand manufacturers. Many of them are familiar models, with a wireless NIC added.

In addition, a number of printer manufacturers are creating wireless adapters for a portion of their existing printer stock. As an example, Dell sells a wireless printer adapter for ninety-nine dollars, which is compatible with their Photo All-In-One 924, 944, 964, 962, and 1700/1710 series printers. If you're serious about allowing printing capability from your wireless network, look into options for adapting what you already have, before you look at buying new. To give you an idea of some of the options available, consider the following discription from the Linksys (Cisco) Web site, explaining their Wireless-G Print Server, which looks like an access point, with a small antenna:

> The Linksys Wireless-G PrintServer lets you connect a USB printer directly to your network, eliminating the need to dedicate a PC to print sharing chores. . . . Connect the PrintServer directly to your network by 10/100 Ethernet cable, or wirelessly over 54Mbps Wireless-G (802.11g). Whichever way the PrintServer is attached to your network, both your wireless and wired PCs will have access to it, and the printer it's connected to. . . . The USB port is compatible with USB 1.1 printers, as well as printers that support the new high-speed USB 2.0 specification for even faster throughput. Your data is protected by up to 128-bit WEP encryption, or pre-shared-key WPA.

You get the idea. There are a number of options. Seriously consider how printing jobs will be sent by patrons and received and handled by the printer, as well as how you will organize patrons picking these jobs up. You don't want hundreds of "orphaned" print jobs hanging out at your one wireless-capable printer, while frustrated patrons look fruitlessly for their jobs at your wired printers. Printing capability of any kind should be included in your site survey criteria, as well, so that you're thinking in those terms from the start.

CHOOSING PERIPHERALS

The delineation of staff and patron networks and roaming issues can become important if you plan to use any peripheral wireless equipment for staff tasks. This category can include, but is not limited to:

- handheld barcode scanners which work as data inventory devices,
- laptops and/or wireless headsets for "roaming reference,"
- laptops for staff collection development which give staff the ability to view and update item records in the ILS system while actually in the stacks, or to search for replacement items for ordering during a weeding project,
- laptops and possibly a wireless router for the creation of a mobile computing lab, to be used for instruction of staff or patrons.

All of these are currently in use in libraries—they are not library "science fiction." Would any of these solutions, or some other wireless solution, improve your staff's flexibility and efficiency? Could you improve customer service with the help of wireless?

Barcode scanners are becoming fairly ubiquitous in business inventory applications. Nearly all libraries with an Integrated Library System (ILS) include some sort of barcode scanner or wand during the checkout process. Fewer libraries are doing inventory collection with handheld scanners in the stacks. Nonetheless, these solutions exist and many of them are offered as part of a larger ILS package. Several Web sites, however, price some of these peripherals separately (www.barcodesinc.com or www.barcode discount.com for instance). Obviously, you will want to look at interoperability, not only with your WLAN (what "flavor" of 802.11 is the device using—b, g?), but also with your ILS, so that you can upload the data from the scanner without any middleware device.

PURCHASING LAPTOPS

An increasing number of libraries are choosing to lend out laptops, with good success. In addition, you may be looking to create a mobile lab of some sort, with wireless-capable laptops connecting to a wireless access point plugged into your public network, or directly to your local WLAN. Your best bet is to buy newer laptops, which include an internal wireless network card (wireless NIC).

As an aside, we use the laptop in-house all the time. It is our 5th Internet station. It isn't connected to a printer and doesn't have any other programs on it so it truly is just an Internet station. Some people prefer using the laptop and ask for it; others may start out using it and move to one of the stationary computers once there is one available. Not surprisingly, teenagers love using it because they can go anywhere in the building (it's a small building!) and sit together while they play games.
Joyce Godwin
Library Director
Indianola Public Library
Indianola, Iowa

If you are "stuck" with some older laptops which are not wireless-capable, you can get wireless NICs that can be inserted in a card slot. Figure 4–3 shows what one of these cards (in this instance from Cisco) looks like.

Figure 4–3 Cisco Aironet 802.11a/b/g Wireless Cardbus Adapter.

Figure 4–4 lists a few examples of external NICs from brand-name manufacturers, with their recent prices.

Figure 4–4 Network Interface Cards (NICs).

Make	Supplier-URL	Description	Price (Avg.)
3Com	www.3com.com	OfficeConnect Wireless 11a/b/g PC Card—Network Adapter	$70
Belkin	www.belkin.com	Belkin Wireless Network Notebook Card, 802.11g, b	$35
Buffalo Technology	www.buffalotech.com	AirStation Adapter-G	$55
D-Link	www.dlink.com	DWL G650 AirPlus Xtreme G Wireless CardBus Adapter	$65
Linksys	www.linksys.com	WPC54G Wireless-G Notebook Adapter	$70
Netgear	www.netgear.com	WG511T 108 Mbps Wireless PC Card	$75

External cards are not recommended, especially for laptops being circulated. In fact, given the choice, we'd recommend not having a laptop available at all unless you can get one with an internal NIC.

We have been lending laptops for three semesters now and it has gone quite smoothly. The service has been well received by the students; they do not mind completing the borrower's agreement or handing over their driver's license in addition to their student ID when borrowing a machine.

There is some discussion about modifying the policy to allow laptops to circulate out of the library for short periods of time, like two or three days. This is just a discussion at this point.

Brian T. Johnstone
Public Services Librarian
Bucks County Community College
Newtown, PA

In the March 2003 issue of *Computers in Libraries*, Nancy Allmang wrote an excellent article on her library's process for implementing a wireless laptop loan service as part of a larger wireless project. Her experience, as excerpted in the following "User Experience," presents good items and issues to consider when purchasing wireless devices for circulation.

We knew what we were after—convenience for our customers. A recent survey had shown that, above all, our customers clamored for additional electronic resources. . . . So we decided on a program that would enable library users to borrow equipment to access the wireless LAN. We planned to allow researchers and guests to check out hardware (wireless laptops or wireless PC cards, and Pocket PCs with expansion packs and wireless CompactFlash, or CF, cards) at the circulation desk for 2 hours at a time.

We outlined what this would mean for our customers and staff. Then we listed the wireless equipment we already had:

- Six Dell laptops with built-in wireless capability
- Four wireless LAN PC cards (for laptops that are not wireless)
- Five iPAQ H3970 Pocket PCs (PDAs with Windows-type operating systems)
- Five Socket wireless LAN CF cards
- Three FlyJacket wireless presentation devices (*http://www.lifeview.com.tw/eng/pro%5fia%5ffyjacket.html*)
- One secure locking cabinet

And then we considered the peripherals we might want to consider buying: portable keyboards and pen text scanners. A short time later we purchased expansion packs to fit around the iPAQs to provide slots for the wireless CF LAN cards.

. . . Here's where we spelled out the details for putting our program into action. Policy and Procedures: First of all, our cataloging staff would catalog all hardware using our Sirsi integrated library system, and then affix bar code stickers. We outlined procedures that our circulation staff would then follow:

- Scan equipment bar codes into the Sirsi system. (This would make it easy to track usage of the equipment.)
- Give out a printed quickstart guide with each piece of hardware borrowed.

- Have the user leave a driver's license or passport at the circulation desk.
- Have the user sign a responsibility form with these caveats: Users 1) will not use the equipment for any malicious purpose, 2) understand the (stated) value of the borrowed hardware, 3) will keep it in the library, and 4) agree to cover the equipment's cost if they're unable to return it.
- Circulation staff will briefly show the user how to sign onto the network using the equipment. A user desiring further help will take the device to the reference desk for librarian assistance.

Later, when laptops were returned, circulation staff would put them aside for re-imaging procedures by our technical support people. (Allmang 2003, 20)

Authors' Note—In 2005, she wrote back in to the magazine, and gave an update on their experience with this service:

You may remember an article a while back ("Our Plan for a Wireless Loan Service." CIL vol. 23, # 3, March 2003) about plans of the Research Library of the National Institute of Standards and Technology (NIST) Information Services Division in Gaithersburg, Maryland, to launch a wireless laptop loan program. We wanted to offer six wireless laptops for customers to borrow at the circulation desk to use throughout the library's three floors and on the patio outside. Those with NIST-owned laptops could borrow wireless LAN cards in the same way and request NIST IT support experts to configure them to work on the wireless LAN.

. . . Now that the program's run for over a year we thought we'd let you know how it turned out. The program was generally successful and attracted lots of positive attention to the library. In July 2004, an analysis of the first complete year's wireless usage showed the laptop program had built a small core of big fans across the organization. Laptops were checked out a total of 145 times, and it was clear that those who'd checked them out had come back for more: 81 percent of all laptops checked out were to repeat customers . . . Two "mega-users" borrowed laptops 14 and 17 times during the program's maiden year and the average borrower checked out a wireless laptop three times . . .

Throughout the first year, customers told us they liked the wireless program. They were able to print to a laser printer equipped with a wireless print server, also located in the library. They did comment that they'd like to he [sic] allowed to bring in and use their personal laptops from home on the wireless network, but strict security regulations prohibit this. During the first year's pilot, two users did check out wireless LAN cards and used them successfully to access the network with NIST-owned laptops.

. . . Now in its second year, the program's been promoted from a pilot to a standard library service. The wireless network's encryption has been upgraded from Wired Equivalent Privacy (WEP) to Wi-Fi Protected Access (WPA). With the new, more secure system, customers can log on to library laptops using their own NIST ID and password and save work directly to their desktop accounts. This is a very popular option. Using the new WPA network they also can map over the network to print wirelessly either in the library or in their offices. (Allmang 2005, 52)

ENSURING INTEROPERABILITY

Interoperability is the idea that all wireless components using the same protocol will be able to work together. The Wi-Fi Alliance (www.wi-fi.org), a nonprofit association originally formed in 1999 under the name Wireless Ethernet Compatibility Alliance (WECA) to promote the growth of wireless LANs, also certifies wireless network products for interoperability based upon the 802.11 standard. Official certification of wireless interoperability by the Wi-Fi Alliance began in 2000, and their Web site currently states that over two thousand wireless products from over two hundred member companies have received certification. All wireless products that have been certified by the Wi-Fi Alliance display the logo shown in Figure 4–5.

Figure 4–5   The Wi-Fi Logo.

Ideally, Wi-Fi certification would eliminate worries on the part of consumers that equipment purchased from different OEMs and used in the same wireless network would not function in tandem. Slowly this situation is working to normalize the interoperability of components. In January 2004, Cnet reported that several hundred Wi-Fi products that had already been certified failed the Wi-Fi Alliance's certification test, "a sign that many pieces of wireless equipment on the market are incapable of working as well as users might expect." However, since that time, additional testing has found that newer products are being designed to work together quite effectively.

To avoid interoperability problems, make sure all of the wireless LAN equipment you purchase has been manufactured by the same company. This is a workable solution, but a potentially tricky one for those who wish to (or are required to) shop around for the best deals on equipment. Furthermore, the solution can only extend so far. If your goal is to provide wireless network access for your patrons with their own wireless-enabled laptops, you cannot control the patron's choice of brand of wireless network adapter card. But you do have control over your routers, bridges, and access points, and maintaining brand consistency among these components may be enough.

Because the standards for wireless LANs are constantly evolving, "upgradability" in your wireless equipment is an important issue to consider. Access points should support firmware upgrades. Ask the equipment vendor if the upgrades can be done from a central location, such as a server, which will automatically distribute the upgrade to all other access points on your network.

802.11n—The Future of Wireless? Currently, the fastest WLANs operate at a max of fifty-four Mbps, a good increase from the previous eleven Mbps. However, there is an initiative to come up with technology that can work at one hundred Mbps. The IEEE's 802.11 Task Group N is in charge of coming up with this standard. The standard may be completed as soon as early 2006. However, products supporting this standard may not be available for a few years. It's good to keep in mind, however, as this will put wireless throughput on a similar level to where desktop, wired throughput is now.

WEIGHING MANUFACTURERS AND EVALUATING EQUIPMENT OPTIONS

There are a wide variety of wireless equipment manufacturers. While there appear to be no significant functional differences among the various equipment manufacturers, there have been some reports that components from different OEMs (original equipment manufacturers) will not work as well together as components from the same manufacturer. If you are receiving recommendations from a vendor, have them explain why they chose particular manufacturers' equipment, and ask if they have experience with these components working together well.

SEARCHING THE WI-FI ALLIANCE WEB SITE

If you are doing your own site survey, or simply want to become more versed in what equipment options are available, see the Wi-Fi Alliance Web site. This site has an excellent Manufacturer Search page, which will be demonstrated in Figures 4–6 through 4–9.

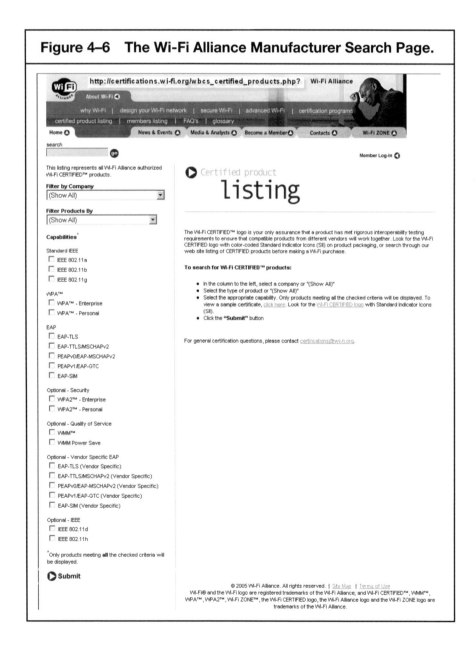

Figure 4–6   The Wi-Fi Alliance Manufacturer Search Page.

If you've already decided on an OEM you can choose to search for a single manufacturer's goods. You can also limit your search to just a single type of equipment (access point, wireless printers, etc.). In Figure 4–7, our search is limited to access points that are 802.11 b, and g capable with WPA or WPA2 security options.

Figure 4–7   Limiting Your Search to One Type of Equipment on Wi-Fi Alliance Web Site.

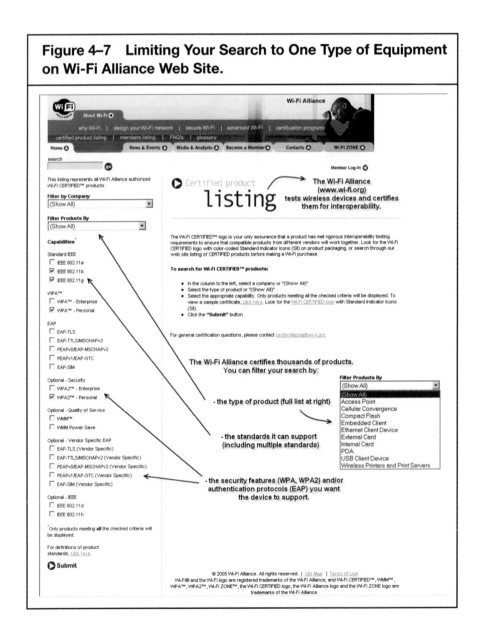

A few of the hundreds of search results are displayed in Figure 4–8.

Figure 4–8   Search Results List on Wi-Fi Alliance Site.

○ Certified product

listing

This is a <u>small</u> sample of the actual search results, which included over 120 products from 35 vendors.

3COM

Access Point

- Office Connect wireless 108Mbps 11g PoE Access Point
- 3Com Managed Access Point

View Wi-Fi Certifications

View Wi-Fi Certifications

3e Technologies International

- 525A-3

View Wi-Fi Certifications

bluesocket 🔊
The leader in secure mobility solutions

- Bluesecure Access Point 1500
- Bluesecure Access Point 1500

View Wi-Fi Certifications

View Wi-Fi Certifications

CISCO SYSTEMS

- Cisco Aironet 1100 Series AIR-AP1121G
- Cisco Aironet 1300 Series AIR-BR1310G
- Cisco Aironet 1200 Series AIR-AP1231G + RM21A 802.11a Module
- Cisco Aironet 1100 Series AIR-AP1131AG
- AIR-AP1231G access point + AIR-RM22A 802.11a radio module
- **Cisco Aironet 1200 Series AIR-AP1232AG Access Point**
- Cisco Aironet 1240 Series AIR-AP1242AG
- Cisco-Linksys WRT54GV5
- Cisco-Linksys WRK54GV2
- **Cisco-Linksys WRTSL54GS**
- Cisco-Linksys WRT54GSV5
- Cisco-Linksys WRTH54GS

View Wi-Fi Certifications

View Wi-Fi Certifications

View Wi-Fi Certifications

View Wi-Fi Certifications

View Wi-Fi Certifications

View Wi-Fi Certifications

View Wi-Fi Certifications

View Wi-Fi Certifications

View Wi-Fi Certifications

View Wi-Fi Certifications

View Wi-Fi Certifications

View Wi-Fi Certifications

D-Link

- High-Speed 2.4GHz Wireless Access Point #DWL-2100AP
- IEEE 802.11g Wireless Access Point/DWL-3200AP
- IEEE 802.11g Wireless Access Point with EoVDSL CPE Modem
- IEEE 802.11g Wireless LAN Office Router

View Wi-Fi Certifications

View Wi-Fi Certifications

View Wi-Fi Certifications

View Wi-Fi Certifications

NETGEAR

- 54Mbps Wireless Access Point/WG602v3
- Netgear 802.11g ProSafe Wireless Access Point-WG-302
- ProSafe 802.11g Wireless Access Point / WG102
- Netgear 802.11g ProSafe Wireless Access Point-WG-302
- FWG114Pv2
- WAG102

View Wi-Fi Certifications

View Wi-Fi Certifications

View Wi-Fi Certifications

View Wi-Fi Certifications

View Wi-Fi Certifications

View Wi-Fi Certifications

As you can see, the list is extensive and includes multiple products from several brand name and lesser-known manufacturers. If you click on the "View Wi-Fi Certifications" link to the right of each product the Wi-Fi Interoperability Certificate appears, indicating the date of certification and what standards the product has passed. (see Figure 4–9)

Figure 4–9   Wi-Fi Interoperability Certificate from Wi-Fi Alliance.

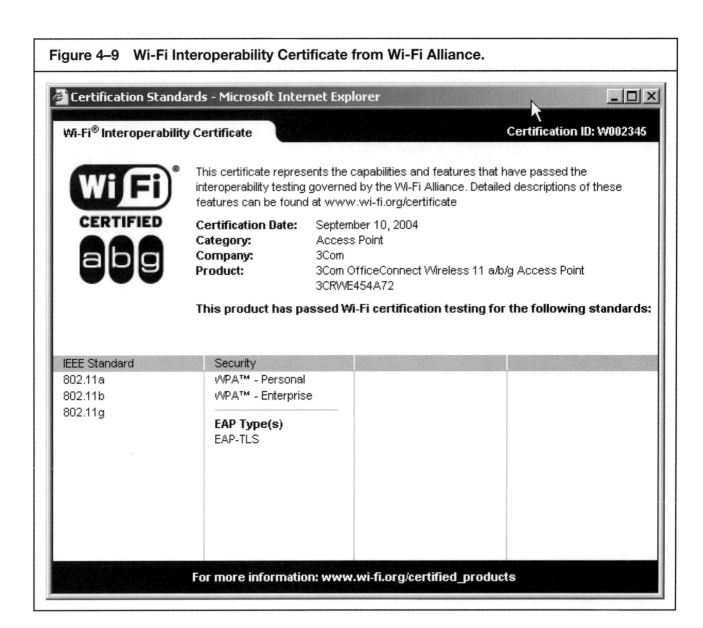

ACCESS POINTS

As explained in Chapter 1, access points (APs) are generally small boxes about the size of a book with a stubby antenna extending from them. They house a transmitter, a receiver, an antenna, and a piece of equipment that acts as a bridge to your wired network. The AP can also function to encrypt and decrypt data streams if your network requires user authentication. The access points transmit data to other APs and also to the end-user's "receiver," which is the wireless network interface card (NIC) installed in their wireless device. Both the AP and the wireless NIC need to use the same standard—IEEE 802.11 a, b, or g—to communicate. These days, APs which can transmit on multiple standards are relatively easy to find, if that is the direction you want to go.

Bandwidth gets divided between those devices receiving from the access point, so to many users this can mean a slower connection. Wireless network access can be slower than direct wired connections to the same source, though the speed difference can vary. The signal from an AP can transmit from one hundred to three hundred feet, but certain high density or reflective structural elements, like elevators or concrete stairwells, can cause interference.

You may want to invest in a different antenna to attach to your APs if you have a difficult building layout. This should be determined in your site survey. Figure 4-10 gives a sample of some access points available from major manufacturers. The prices are current as of January 2006 and reflect a rough price range for this type of device. We would suggest using the Wi-Fi Alliance site, as described earlier, for a more comprehensive equipment search.

Figure 4–10 Sample Recommended Wireless Access Points and Routers.

Make	Supplier-URL	Description	Price (Avg.)
AsusTeK	www.asus.com	WL–330 Access Point	$55
AsusTeK	www.asus.com	WL–500g Wireless Router	$85
Buffalo Technology	www.buffalotech.com	AirStation 54 Mbps Wireless Broadband Router	$90
Cisco	www.cisco.com	Aironet	
D-Link	www.dlink.com	DI–524 AirPlus G Router	$50
D-Link	www.dlink.com	DWL–7100AP Wireless Access Point with SNMP	$150
HP	www.hp.com	ProCurve Wireless Access Point 420 NA	$350
Linksys, Inc.	www.linksys.com	WRT54G Wireless-G Broadband Router	$60
Linksys, Inc.	www.linksys.com	WRT55AG Wireless A+ G Broadband Router	$96
Netgear	www.netgear.com	WGR614 Cable/DSL wireless router (54 Mbps/2.4 GHz)	$70
Netgear	www.netgear.com	WG302 802.11g ProSafe Wireless Access Point	$250
SMC	www.smc.com	Barricade 7004VWBR Wireless Router	$60
SMC	www.smc.com	EliteConnect SMC2552W-G 2.4 GHz 802.11g Wireless Access Point	$305
U.S. Robotics	www.usrobotics.com	USR8054 802.11g Wireless Turbo Router	$70

WIRELESS BRIDGES

Several vendors offer what they refer to as wireless LAN bridges, which are somewhat different from access points, and are not clearly defined in 802.11. The industry-accepted definition of a bridge is a device that connects two networks that may use the same or a different Data Link Layer protocol. Bridges have been in use in wired networks for some time, used to connect two Ethernet LANs, for instance. They can also be used in the wireless environment to extend the range of an existing WLAN and connect devices that don't have wireless NICs into a WLAN.

LAN bridges have ports that connect two or more separate networks. The bridge receives packets on one port and re-transmits them on another port, once it recieves its packet. In this way, stations on either side of a bridge can transmit packets without collisions. Some bridges retransmit every packet regardless of destination. A learning bridge, on the other hand examines the destination address of every packet to determine whether it should forward the packet based on a decision table that the bridge builds over time.

Bridges vs. Access Points

Access points connect multiple users on a wireless LAN to each other and to a wired network. For example, twenty users equipped with wireless NICs may associate with a single access point that connects to a wired Ethernet network. Each of these users has access to the Ethernet network and to each other.

The access point here works as a bridge device, but the access point interfaces a network to multiple users, not to other networks. According to Jim Geier in an article for Wi-Fi Planet.com bridges . . . connect networks and are often less expensive than access points. For example, a wireless LAN bridge can interface an Ethernet network directly to a particular access point. This may be necessary if you have a few devices, possibly in a far reaching part of the facility, that are interconnected via Ethernet. A wireless LAN bridge plugs into this Ethernet network and uses the 802.11 protocol to communicate with an access point that's within range. In this manner, a bridge enables you to wirelessly connect a cluster of users (actually a network) to an access point. (Geir 2003)

Types of WLAN Bridges

- *Basic Ethernet-to-Wireless* Bridges connect directly to a single device via an Ethernet port, and then transmit a wireless signal to an access point. This can be useful in older devices where adding a wireless NIC, is impractical.

- *Workgroup Bridges* connect wireless networks to larger, wired Ethernet networks. According to Geier, "a workgroup bridge acts as a wireless client on the wireless LAN and then interfaces to a wired network. The wired side may connect directly with a single device (like an Ethernet-to-Wireless bridge) or to an Ethernet hub or switch that connects multiple devices. Generally, a workgroup bridge offers higher-end management and security utilities (with higher prices) as compared to a basic bridge." (Geir, 2003).

- *Access Point / Wireless Bridge Combos* are access points that can be configured as bridge, but not both at the same time. Linksys Cisco, and Proxim, among others, offers such devices. These access points can operate in point-to-point and point-to-multipoint bridge mode.

Antennae

Your site surveyor may recommend that you buy upgraded antennae for some of your access points or wireless bridges. This may be due to interference issues in your building, negative experience they have had with that type of equipment in your type of building, or a special need, such as the ability to send and receive between buildings. It is unlikely in a smaller organization or building that you will need to make many antennae changes. One change you might need to make, however, is to buy directional instead of omnidirectional antennae. Directional antennae are particularly helpful if you are trying to send signals down a long narrow corridor or space, or if you are mounting your AP on a wall and want the signal directed only out into the room, not drifting to outer areas. Your site survey should include at least a rough map of the expected signal range and direction for each AP or bridge, so you can understand why a different antennae might be needed.

In their *802.11 (Wi-Fi) Networking Handbook*, Neil Reid and Ron Seide warn those planning to mount antennas above ceiling tiles to check the plenum rating for the space above the tiles. The **plenum** is the area above a room where heating and air conditioning run (Reid and Seide 2003, 163). Any device placed in the plenum area must meet the UL2043 standard for fire safety. Make sure you, your vendor or your IT crew are aware of this. This also comes into play if you are running electrical access to an area so you can plug in an AP. For this situation we highly rec-

---

Plenum
The area above a room where heating and air conditioning run. Any device placed in the plenum area must meet the UL2043 standard for fire safety.

ommend that you have a professional electrician do this type of work. If you are having network cabling run to a distant or inaccessible access point, you will also want to be aware of where it is being run, and whether it meets code. Any professional site surveyor should be aware of these issues and can answer your questions. If you're doing the cabling yourself, talk to a local IT vendor to be sure of what you need.

CONSIDERING WLAN MANAGEMENT OPTIONS

Okay, you've got your site survey, you've got your implantation plan, you know where all the equipment is going to go. Now . . . how are you going to manage it all? How will you restart access points without having to crawl into your ceiling? How will you even know those access points are malfunctioning, besides the whining from disconnected patrons? Your plan prohibits use of telnet, FTP, and streaming content. You also want to limit bandwidth, so a certain number of users can access one AP at a time, to manage usage. How do you implement these policies?

Here we divert a bit from "hardware" per se, and talk about software options for managing all of this hardware. In the library world, there are several ILS vendors and vendors of print and time management products for public access computers who are now extending their management systems to include any wireless networks in your system. One of the greatest advantages of wireless network management options is the ability to monitor the wireless access points and attached devices. In particular, the ability to reset APs remotely when they "go out," as they sometimes do.

Our wireless network is handled by our city IT staff, with whom I have a good relationship. However, all monitoring and maintenance of our public network takes place next door at City Hall. After I and my colleagues complained that we had no concrete way to respond to occasional patron complaints that "the wireless isn't working," the IT guys got me a little device from Targus (Targus Wi-Fi Scanner Model ACW20 from www.targus.com) that lets me press a button and check, on the fly, if the SSID is broadcasting and what signal strength is showing at a particular spot in the library. However, it does not tell me how many users are connected, if they're getting through, or which access point of our four might not be working—though by wandering around, I can sometimes figure it out and call to have it reset. (Usually I just call and let them figure out which one needs resetting.)

I have to admit that my little toy is most useful for proving to patrons that the library does indeed have wireless signal, and helps me convince them to go back to their laptop and try the extensive list of troubleshooting options I've given them on a handout!

Louise Alcorn
Reference Technology Librarian
West Des Moines Public Library
West Des Moines, Iowa

We have listed a number of vendors in Source B, but one vendor we heard very good things about from several librarians was ILS vendor Polaris. Their WAM (Wireless Access Manager) system works with a number of ILS types using the SIP protocol. With it, you can limit the times the WLAN is available, meter bandwidth, restrict protocols like FTP and telnet, and produce usage reports. Here's one library's experience with this product:

The Polaris WAM product has been excellent to use. It is "ILS agnostic" so will work with any ILS using SIP2. In our initial wireless project, we bought a whole package for our libraries that included a Linksys AP, a D-Link 3100 gateway product and the Polaris software which we put on our SIP server. We require all users to be library card holders, and authenticate them through SIP by library card number. We had a little trouble early on with a proxy server sitting on another subnet—it was making the end user make changes to settings so it couldn't connect properly. Once we brought that server into this subnet, it worked just fine.

One of the best aspects of the Polaris WAM product is the ability to track wireless usage in all of our libraries. The accounting piece of all this is that I use SQL server to track wireless sessions in the same way that you track usage of items and wired sessions. When we talk about accountability to our board, they ask, "Vicki, you want to put up a wireless system, but if we run our libraries like we should, like a business, what does it mean to offer wireless access? We keep a gate count, and item use counts, why shouldn't we also track use of our wireless (and

wired) networks?" This product allows us to do that, and fairly seamlessly. We are about to update our entire district system to a Cisco-based solution that will give us both data and voice telephony capabilities, including wireless telephony. We will continue to use the Polaris WAM product on top of that to authenticate and track usage. WAM has been very flexible—more flexible than we dreamed it would be!
Vicki Terbovich
Chief Technology Officer
Maricopa County Library District
Phoenix, AZ

MANAGING INSTALLATION

We've already discussed many of the issues involved in installation, including clarifying where equipment will be placed and planning ahead for cabling and electrical runs, as needed. With placement and installation of access points, you should consider how to balance visibility and accessibility. You don't necessarily want them out where patrons can handle them, or even see them. Whether or not you are installing a remote management system, be sure that technical staff can get to the APs without undue difficulty. If you have a drop ceiling with easily removable tiles, you might place them in this upper space, assuming you can run cable and power there.

Regarding power, we have mentioned before the option of using Power Over Ethernet (PoE) adapters for difficult to access areas. These are fairly inexpensive—one model, the DWL-P200 from D-Link, retails at about $44.99. It transmits both power and data to Ethernet-enabled devices up to one hundred meters, which is the maximum recommended length for Ethernet cable before attenuation begins. A very negligible amount of throughput is given up using these devices. Check the power needs of the device (AP, switch) you're plugging in carefully, to be sure the PoE adapter can handle it. Most basic APs' power output can be handled easily by such devices.

If you are working with an outside vendor, be sure their project plan includes all aspects of installation, including cabling and electrical, even if they're sub-contracting (or recommending that you sub-contract) that portion of the installation. You don't want them showing up and saying, "We can't put this here—there's no power!" or "The cable won't reach to here, and if we put it over there, you won't get signal in your café." This may seem obvious, but in the throes of a technology project, noting such minutiae can save you major headaches later.

THINKING OUTSIDE THE BUILDING: SATELLITE INTERNET AND WIRELESS

Use your imagination to think of other ways that wireless can help you, beyond expanding your network capabilities to be accessible via wireless in your building(s). For example, some libraries are using wireless on their bookmobiles. If you've ever taken a cruise and used an onboard computer to access the Internet, you've used satellite-transmitted Internet service. Some libraries, including the Maricopa County Library District, use satellite Internet to connect stations in their bookmobiles. Once the bookmobile stops and sets up, they connect to the satellite service and provide real-time library services from this "moveable library."

Wireless Technology on Board the Bookmobile

The Maricopa County Library District uses wireless technology aboard its thirty-two foot bookmobile enabling staff to provide library services to its customers in real time. Staff are able to process materials (check in and out) and place requests in the system. This provides customers with accurate and current account information as well as the status of their requested materials. They are also able to use all of the library's electronic resources available through the library Web site.

Staff use laptops to connect via a WAP to the server which is housed in a cabinet at the rear of the bookmobile. Through VNCViewer (software that allows them to view and access the satellite server), they can raise and lower the satellite dish. Once the dish is in place and staff connect to the Internet they then use Terminal Server to connect to the library's online computer system. Wireless barcode scanners also play a part in this truly innovative way in which customers are served remotely using wireless technology.

Paula Wilson
Web Services
Maricopa County Library District
Phoenix, AZ

This satellite service was not cheap to implement. According to Paula's colleague, Vicki Terbovich, Chief Technology Officer at MCLD, the bookmobile system cost about ten thousand dollars per unit to install. The remarkable service they're able to provide on the bookmobiles makes this very much worth it, however. They can connect in real time via terminal server to their Polaris ILS and can check out books and place holds in real time, rather than having to take everything back to the home library and input the data. They also have laptops on the vehicle so patrons in remote areas can access the Internet. We're all aware of the difficulties of bringing Internet service to many rural areas, so you can see the boon these mobile labs would be for remote patrons.

MAKING EQUIPMENT WORK FOR YOU

Equipment choice is very dependent on what you want wireless to do for you. In fact, the equipment should be a slave to your overall plan, not the other way around. Marshall Breeding offers the following perspective:

> Given that there are technical solutions to a very broad range of options for implementing a wireless network, the shape of a library's wireless network service can be formed through an administrative process; it need not be dictated by technical or security requirements. If the library administrators, for example, elect to offer a convenient, user-friendly open wireless network, the library's technical staff should be able to design and implement such a service without compromising the security of the library's overall network. If, on the other hand, library management chooses an authenticated and encrypted network, technical solutions are readily available. (Breeding 2005, 33)

As you've heard from the many librarians who have shared their experiences throughout this book, wireless implementation can run from the very simple—a single access point hooked into an outside DSL line, providing Internet access for patrons—to the very complex and integrated, where multiple floors, multiple buildings, older buildings, staff and patron needs, and even moving bookmobiles can come into play. Equipment options exist for all of these possibilities. Think creatively, ask questions, do research, and ultimately decide what is best for your organization in terms of service, scope, and budget, then buy accordingly. Source D in The Wireless Sourcebook provides you with some contact information for some of the larger manufacturers. We also recommend highly a look at the Wi-Fi Alliance Web site (www.wi-fi.org) for your initial equipment research.

5 SECURING, MAINTAINING, AND TROUBLESHOOTING WIRELESS NETWORKS

OVERVIEW
• Establishing Wireless Security • Wireless Security Options • Assuming Responsibility for Security • Maintenance and Troubleshooting

---

ESTABLISHING WIRELESS SECURITY

Many organizations hesitate to implement wireless networks because they've heard that such networks are terribly insecure and will open them up to security violations. In the early days of wireless data networking, this was a reasonable position. However, as wireless has evolved, so have the available security options. In some cases, they can now rival wired networks in security. Let's look at some of the concerns regarding wireless networks and some solutions for securing them.

In his *ALA Library Technology Report*, "Wireless Networks in Libraries," Marshall Breeding comments:

> In the early days of wireless networks, it was fairly common for an organization to have wireless networks set up in such a way that left the entire network vulnerable. For instance, so as not to have to wait for an overworked IT staff to add additional network connections, non-IT departments (without the support of IT personnel) often set up rogue WLANs, yielding an informal network with wireless Internet access for the department. These WLANs almost always had the security features disabled and were often positioned in such a way that could provide inappropriate access to the organization's business systems. (Breeding 2005, 25)

These networks were being created with an eye more toward

War driving
Typically refers to driving around with a wireless-enabled laptop and antenna to locate and exploit security-exposed WLANs.

War chalking
Marking buildings or sidewalks with chalk to show others where it's possible to access an exposed company wireless network. These access points are typically found through war driving.

Key
In encryption, a key specifies the particular transformation of plaintext into ciphertext, or vice versa during decryption. Keys are also used in other cryptographic algorithms used for authentication.

expediency and access rather than to the security of data or users. This resulted in the amusingly named but harmful practice of **war driving**. This is simply the practice of driving around with a laptop having a powerful antenna, searching for accessible, vulnerable wireless networks. This then led to **war chalking**, the practice of chalking graffiti near the hacked site telling other freeloaders where they could access the network, and its characteristics, including its SSID (its unique name, needed for accessing) and which version of 802.11 standards it was using. Before you start checking your library's parking lot for chalk marks, keep in mind that the vast majority of business wireless networks today have potentially multiple levels of security, including heavy-duty encryption. Because free or low-fee Wi-Fi for simple Internet access (one common goal of war driving) is so readily available, this sort of behavior has become pointless.

EAVESDROPPING AND ENCRYPTION

Concerns about "eavesdropping" are reasonable, when using a wired network. Neither Ethernet's 802.3 CSMA/CD nor 802.11a/b/g CSMA/CA "preclude the possibility that one network station will be able to open and read the packets of its neighbor's station." (Breeding 2005, 24) Eavesdropping, by means of software that can "sniff" for data being carried on the network, can endanger both the privacy of users connected to the network, and the security of any data stored on the network. In a wired environment, this might require the intruder to have a physical connection to the network, such as plugging their laptop into a live Ethernet port. In a wireless environment, this is trickier due to the need only for simple proximity to an insecure access point and the prevalence of clients using their own equipment which could be loaded with harmful software.

The basic solution to eavesdropping and unauthorized access is to use encryption. Encrypted data is encrypted before it hits the network and can only be decrypted by providing pre-specified information, usually in the form of a **key** and other data. A key is a string of characters, of varying lengths, depending on the scheme used. The trick is making sure that the key isn't readily available to the outside intruder, while also providing that key to any devices you do want to connect to the network. This is true for both wired and wireless networks. In the case of a wired patron network, for example, this is done fairly seamlessly without much user intervention, as each machine on your network is controlled by you. When you're talking about outside machines connecting to a wireless network, however, this is more complicated, as you need a way to provide the key to new users. We'll discuss some encryption schemes for wireless networks later, but first let's look at the overall design of your network(s).

THE SEPARATION OF STAFF AND PATRON NETWORKS

To a great extent, the security of your wireless network is dependent on the security of your wired network. The first question you must ask is whether your staff and patron networks are separated. Not having them separated exposes you to dangerous insecurities. They can be physically separated, using separate switches or hub ports, or may simply use a VLAN (Virtual Local Area Network) arrangement so the staff network's data is not viewable to patron network users. You may also have chosen, as many libraries have, to put your patron network on the outside of one or more firewalls, or at least in the "DMZ" (demilitarized zone—a semi-controlled interim "space"), where you assume some amount of deliberate or accidental mischief is possible and where it can be controlled.

Why is this separation important? There are two reasons. First, as a library entity, chances are that you have a set of regulations, possibly even state laws, which require that your use of library patrons' personal information (names, addresses, telephone numbers, student ID numbers) be carefully controlled. This information is likely stored on your ILS (integrated library system), which may or may not be separated, but may also be stored in the form of usage reports, overdue notices, etc., which may reside openly on your staff network. Access to this information by outside, unauthorized users could put your library and by extension, your university, company, or municipality, in a world of trouble.

The second reason keeping staff and patron networks separated is important is that such an arrangement makes the implementation of wireless extensions to the network that much easier. If you are planning on wireless for the public only, you can simply make the wireless an extension of the public network, so that providing open access, as you may want to do, will not endanger your staff network. The security that protects your staff network from mischief on your patron network would also protect it from the WLAN extension to that patron network. In the same way, if you decide you need a staff-side WLAN, you can decide whether they will simply use the public WLAN—reasonable if you simply want them to have Internet access at some remote location—or if you will have a secure WLAN that provides staff access to information on the staff network.

AUTHENTICATION

In most enterprise networks, some form of authentication is used. While encryption protects the data between devices so it is not being sent as simple text accessible to anyone looking for it, authentication is based on the idea that in enterprise networks, you have a large number of users that need to be authenticated and authorized before accessing a network's resources. No device keeps records of all the usernames, passwords, authorization credentials, etc. Instead, larger organizations use an authentication server, which stores this information and can be queried when users attempt to connect. Authentication servers generally follow one of two standards: RADIUS (remote authentication dial-in user service) or LDAP (lightweight directory access protocol). Authentication is one measure of controlling access to your resources and securing your network. Let's discuss some of the other options for securing a wireless network.

WIRELESS SECURITY OPTIONS

Most access points come with the ability to provide basic encryption security, but come with that option turned off. In the simplest possible wireless provision option, if you have a completely separate line out to the Internet—a broadband connection provided by a local vendor, for instance, or a separate T–1 or DSL line for the purpose—you could plug your access point into that line and with almost no effort, provide free, open access to the Internet for anyone in range with a wireless receiver. Does this sound terrifying? Why? It doesn't impact your network at all, and if it's provided by an outside ISP or vendor under a goodwill program, it probably costs you nothing. Issues might arise if you normally provide content-filtered access to the Internet in your building, but that's a policy issue you need to address before you start providing wireless. You could still use the same open setup, but work with the provider to run the access through a network-based filter program.

If you decide to secure your network, your first layer of protection is an encryption scheme. There are several options available for WLANs, which have evolved with the growth of wireless data networks.

WEP: WIRED EQUIVALENT PRIVACY

WEP (wired-equivalent privacy)
Wired-Equivalent Privacy protocol was specified in the IEEE 802.11 standard to provide a WLAN with a minimal level of security and privacy comparable to a typical wired LAN, using data encryption. It's now widely recognized as flawed because of an insufficient key length and other problems, and can be cracked in a short time with readily available tools.

The first version of WLAN security protocols was called **WEP** (**wired equivalent privacy**). The name is a misnomer, as it was meant to suggest that you were offered privacy controls similar to a wired network, to keep eavesdropping at bay. WEP is still available as an option on most access points, though APs usually come with any security disabled. It's a simple setup—you load the software, choose the security option, create an administrative password, then have it create random WEP key(s). WEP matches the key with a twenty-four bit initialization vector (IV), which is then transmitted in the packet so the receiving device, as long as it has the WEP key, can decrypt and connect. WEP also makes use of a thirty-two bit encryption method, or "integrity check value," which is sent along with the packet to verify its authenticity. Thirty-two bit is considered fairly low-level security at this point.

In order for the receiving device (e.g. a patron's laptop) to work, you need to provide any device accessing the access point with the key. This is where it gets problematic for libraries. You would have to offer a key to any user who wanted to connect to the network. In many libraries, to be provided a key patrons have to come to a service desk and "register" in some fashion for the service. This means a lot of staff time, especially if you change the key from time to time, which is recommended for increased security.

We didn't want the wireless users to have access to our network so found a local Internet provider who is currently providing access for free. The wireless access is therefore completely separate from our other computers. We purchased the router, installed an electrical outlet, bought a laptop, got it all installed and that was it!

We started out fairly restrictive. The wireless router "allows" us to specify the hours that the connection is available so we have it available only during library hours. This satisfied any qualms the board and staff had about people surfing the Internet and doing illegal things from the Library parking lot. We also changed the WEP key (we refer to it as the password) on a weekly basis. People have to stop at the desk and pick up a card to logon. Because librarians keep track of everything, I count how many cards are used each time to come up with a rough estimate of how many times the wireless connection has been used by patrons coming into the library.

A year later, we're a little bit looser. I change the WEP key monthly now and don't care so much about how many patrons we have had use it. But we still only have it available during open library hours.
Joyce Godwin
Library Director
Indianola Public Library
Indianola, Iowa

WEP has several problems, the most serious of which is that it is easily cracked. It has relatively short keys, and there are a number of products out there now, with charming names like WEPCrack and AirSnort, which can pull a small amount of traffic and find the WEP key from that data. In addition, since you manually have to assign new keys with WEP, "on the fly" securing is cumbersome.

Keep in mind that some security is better than no security at all, if you don't want a fully open network. For a home network, for instance, unless you're running a business from your home or are frequently sending sensitive information to/from your employer, a simple WEP system would be adequate. Even for a small library where the WLAN is quite separate and you have the need to pre-register users (for some needed metric or to require them to sign a user agreement), this system would give you a basic measure of security that may very well be sufficient.

802.11 AND WPA WI-FI PROTECTED ACCESS

Obviously this minimal security would not be sufficient for enterprise wireless networks. A more rigorous approach would be needed. The IEEE's 802.11 group began to develop the 802.11i standard, which specifies a security framework for WLANs. Even before ratification of 802.11i, which didn't happen until June 2004, versions of these improved requirements were being developed. The first phase of these came out as **WPA**, or Wi-Fi Protected Access, which plugs several of the "holes" left open by WEP.

WPA specifies a key length of 128 bits, which is much harder to break. WPA also uses a protocol called **TKIP** (Temporal Key Integrity Protocol). As mentioned earlier, one of the problems with WEP is that it uses keys that are not changed, or changed only infrequently since they require manual changing. WPA also uses a mix of the key with the IV (initialization vector), but it uses a forty-eight bit IV. According to Wikipedia.org:

WPA (WiFi Protected Access)
Wi-Fi protected access is a data encryption specification for 802.11 wireless networks that replaces the weaker WEP. Created by the Wi-Fi Alliance before the 802.11i security standard was ratified by the IEEE, it improves on WEP by using dynamic keys, extensible authentication protocol (EAP) to secure network access, and an encryption method called temporal key integrity protocol (TKIP) to secure data transmissions.

TKIP (Temporal Key Integrity Protocol)
WPA corrected some deficiencies in the older Wired Equivalent Privacy (WEP) standard, including adding the TKIP security protocol. TKIP hashes the initialization vector (IV) values with the WPA key to form the identifying traffic key, addressing one of WEP's largest security weaknesses. WEP simply concatenated its key with the IV, which was less complex, and easier to break. With TKIP, every data packet is sent with its own unique encryption key, so is more secure.

TKIP hashes the initialization vector (IV) values, which are sent as plaintext, with the WPA key to form the RC4 traffic key, addressing one of WEP's largest security weaknesses. WEP simply concatenated its key with the IV to form the traffic key, allowing a successful related key attack. The practical upshot: TKIP ensures that every data packet is sent with its own unique encryption key and is, as a result, much harder to crack.

Earlier we discussed authentication schemes for networks in general. WPA and 802.11i's security protocols incorporate what is known as 802.1x authentication (see sidebar). This uses the RADIUS or LDAP schemes described earlier to perform a request for authentication as a requirement for the user's device to connect to an access point. Basically, the user's device requests authentication to a device, usually an access point, that requires authentication. An authentication server manages the interaction, using either RADIUS or LDAP. This is managed using a language called **Extensible Authentication Protocol (EAP)**. If the access point requires 802.1x authentication, it uses EAP to verify before letting a user's device connect. In addition, plugging another WEP hole, the authentication server will also distribute dynamically generated keys to users' devices, and periodically refresh them. Marshall Breeding notes:

> 802.1x works well in organizations that already have an infrastructure that includes LDAP or RADIUS. For those that don't WPA supports the use of pre-shared keys, allowing a secret key to be manually distributed to each authorized client. Although not as secure as using 802.1x, WPA with pre-shared keys is much more secure than WEP. (Breeding 2005, 28)

Windows XP's wireless networking capabilities support WPA, and include the option for this level of authentication while most access points being made now also support WPA. Check the product specifications to be sure. WPA was designed to tighten WEP but not require major hardware upgrades, and most equipment designed for WEP was able to support WPA by upgrading software or firmware only.

EAP (Extensible Authentication Protocol)

According to Wikipedia.org, "EAP (pronounced 'eep') is a universal authentication mechanism, frequently used in wireless networks and point-to-point connections. Although the EAP protocol is not limited to wireless LAN networks and can be used for wired LAN authentication, it is most often used in wireless LAN networks. Recently, the WPA and WPA2 standard has officially adopted five EAP types as its official authentication mechanisms."

Troubleshooting note: If a user has 802.1x "turned on" in his wireless options, and you have an open network that does not require authentication, he may connect at first, but keep "blipping out" every few minutes, as his laptop tries to authenticate to your system. Have him check that this option is turned off. He should see it in his Wireless Network Connection Properties→Wireless Networks tab→choose your library's network SSID from the list, then Properties→Authentication tab. It will rarely be on if your network does not require data encryption, except if it was previously configured for static, secured access to an employer or university system. Uncheck the box to disable it (see Figure 5–1).

Figure 5–1   802.1x Security Troubleshooting.

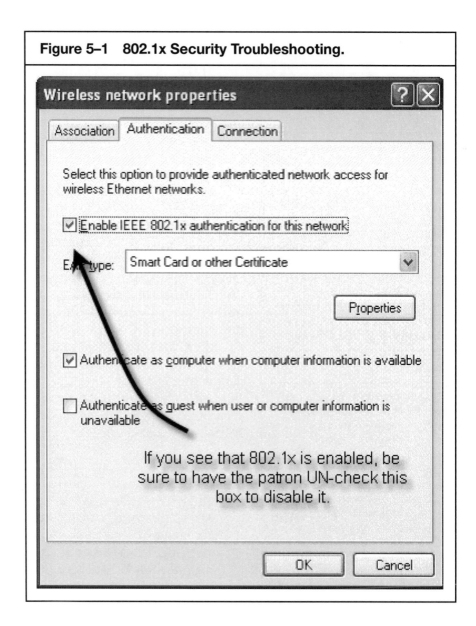

WPA2

Wi-Fi protected access 2 is an enhanced version of WPA. It is the official 802.11i standard that was ratified by the IEEE in June 2004. It uses advanced encryption standard instead of TKIP (see **WPA**), which supports 128-bit, 192-bit and 256-bit keys. The WPA logo certifies that devices are compliant with a subset of the IEEE 802.11i protocol. WPA2 certifies full support for 802.11i. WPA and WPA2 use a key hierarchy that generates new encryption keys each time a mobile device establishes itself with an access point. Like WEP, keys can still be entered manually (preshared keys); however, using a RADIUS authentication server provides automatic key generation and enterprise-wide authentication.

WPA2

With the ratification of the 802.11i standard, this specification is being called "**WPA2.**" One upgrade to WPA that comes with the WPA2 protocol is the use of the Advanced Encryption Standard (AES), which is similar in function to TKIP but is much stronger. Unlike WPA, which was designed to make a network more secure without hardware changes, however, WPA2 is not compatible with WEP and may require a hardware upgrade. Again, check product specifications to see what is supported. You may even find that WPA is sufficient for your security needs and you don't require WPA2.

If you do not have an authentication server or an 802.1x capable structure but you still want the stronger levels of security like WPA and WPA2, you can use some commercial Wi-Fi security services. One example is the SecureMyWiFi product from WiTopia (www.witopia.net/aboutsecuremy.html) which has the ability to dynamically change encryption keys, limit use to preregistered devices, and also offers a secure, Web-based remote monitoring and management tool. Considering that WPA and WPA2 are fairly new, keep your eye open for additional products that will likely be coming soon.

VPN: VIRTUAL PRIVATE NETWORKS

The concept of the VPN (virtual private networks) predates wireless data networks, but has applications in a wireless environment. A VPN creates a "tunnel," or secure channel of communication that allows information to travel safely through insecure networks. Software on both ends of the transmission encrypts and decrypts the data. In comparison, encryption on a WLAN is on the data traffic between the laptop, or end-user radio device, and the access point only. Data is not encrypted out on the Internet, for instance, or at the remote host if a user is accessing an Internet site (most secure Web sites do provide Secure Socket Layer [SSL] protection, at their end).

An example of VPN usage would be employees who telecommute and may want to access data on their employer's network. Louise sees them every day in her library, using her library's wireless hot spot. If a company has set up a VPN, it will have software loaded on its receiving server which will also be loaded on the employees' laptops, allowing them to remotely access information at the employer's location. If a VPN is set up, even an open WLAN can be used without too much concern for proprietary or private data.

We just started wireless access last fall . . . in our library. It is available only to current students, staff and faculty. There is no guest access. Computers connecting to the network must have Cisco VPN software installed, and at present there is no way to get the software other than by going to the tech support station in the library. Logging in requires that a user know their username (something called "Standard UM-M Computer Access User ID" or SCAUID for short) that they get by logging into a different Web site with a different login name and password.

Basically, our wireless network here is very secure but involves a complex and confusing login process and extra software. Oh, and for a while the software was only available for PC. Now we can also support Mac and Linux though, so there's been some progress. But we still spend most of our time re: wireless just troubleshooting software and login problems. It's not the seamless process that most students, staff and faculty think of when they think "wireless Internet," and it has caused a lot of frustration.

Samantha Schmehl Hines
Outreach Coordinator/Social Sciences Librarian
Assistant Professor
Mansfield Library
University of Montana

ASSUMING RESPONSIBILITY FOR SECURITY

Where does the responsibility for security lie in a library wireless environment? A library's primary security responsibility is to secure its own institution's system and data. In any open hot spot, users must assume some responsibility for their own equipment. In particular, they need to realize that the environment may not be secure, so that transmitting private information should be handled with care, and virus and spyware protections should be kept up to date. In the previous VPN example, if employees are doing work on laptops approved or provided by an employer, secure access to work resources, should be provided by that employer. It should not be assumed that a library where the employee will work will have secure access.

As libraries are service-oriented environments, we tend to balk at leaving a user "out in the cold." There are easy ways to achieve some balance, however. Secure your environment as you feel is appropriate for your library, whether it be fully open, keys handed out manually after registration, or full authentication needed (possibly connected to an existing authentication scheme with a university network). Do whatever makes sense for your situation and maintains your policies on information privacy and also on public network use. Facilitate patrons' use of this system by creating helpful documents and FAQs. You may also want to make use of a captive portal to communicate your user agreement. Be aware, however, that users rarely read these "click-through" agreements, so don't put anything of vital importance there that you don't repeat elsewhere.

WEP is now thoroughly discredited, but 802.1x and 802.11i still don't have universal support. WPA remains an interim standard, but doesn't appear to offer much security for public installations. In those cases, it appears "captive portals" are still the best way to go.

. . . [Our wireless system is] configured as a captive portal, so patrons are presented with a login page as they try to browse the Web. University students, faculty, and staff use their university computer account username/password. Theoretically we could accommodate non-member patrons with a button that asks them to accept the terms of service or a form to enter billing information, but we've not explored that yet. Hardware needs are minimal, requiring only that patrons have 802.11b/g compatible network cards. We [the university's IT folks] just implemented a "pre-clearing" program where the Shasta device checks that current virus protection and updates are running on the computer before it will allow the connection. We're only enforcing this for Windows-based devices, and we're giving away the virus protection software to all students, faculty, and staff.

For those with access credentials and up-to-date software, it's as easy to use as Panera's increasingly ubiquitous free Wi-Fi service.

Casey Bisson
Library Information Technologist
Lamson Library
Plymouth State University
Plymouth, New Hampshire

MAINTENANCE AND TROUBLESHOOTING

While setting up and maintaining a wireless network bring with it some minor pitfalls, most network administrators will find that wireless networks are somewhat easier than their wired counterparts in terms of upkeep and troubleshooting. There are some things you can do to make your maintenance and troubleshooting job easier, however.

Once a wireless LAN is set up and running, it usually involves few problems. However, there may be instances during the initial installation and setup where issues may arise. Maintaining and troubleshooting the WLAN itself is largely a matter of making sure that access points are not moved or damaged, that signal is getting to areas you want covered, and that any firewalls or security you've set up are continuing to work properly. Basically these are the normal network maintenance issues without all the cabling and client computer support issues. Instead, you have to deal with making sure that signal is transmitting as you have planned, as well as some unique pitfalls that come with having user-owned equipment connect to your network (e.g., patrons not knowing how their own laptops work, and assuming that you do!). Once your site survey is done as described in Chapter 3, you will still likely require some additional "tweaking" of access point locations to make sure signal reaches all coverage areas. This is largely a matter of trial and error and a lot of roaming around with a wireless-enabled laptop to "catch" signal and measure signal strength. There are more sophisticated tools to review signal, but since your users will be using their laptops to find your service, knowing how they'll be receiving this signal is as useful as anything.

COMMON PROBLEMS

We'd love to say that a WLAN runs itself, but that would be misleading. Basic wireless data networks are remarkably easy to install, and on the whole don't cause huge numbers of problems, but they are not without issues. You will have some physical connections, especially the primary connection to your wired backbone network, and you may also have to deal with interference to your signal-connected equipment (from a wireless switch to a distant access point, for instance). The most common problem is patrons complaining that their access would "blip out" from time to time, sometimes chronically. This could be the access points downing, or it could be user error with their own equipment. If you have monitoring software in place such as Polaris' WAMS product, you would have remote access to your WLAN, and could reset the access points as needed. In a small library with just a few access points, you may be able to physically turn the APs on and off, which resets them. Patrons should be able to reconnect at that point, assuming their device is properly configured (see "Troubleshooting for Users" below).

The Roosevelt University library was included in the campus-wide setup of wireless access points circa 2002. The Schaumburg Campus Library's node is situated above the ceiling tiles in the Main Room. Periodically, and for unknown reasons (microwave? cell phones?) the wireless node will lose its connection and requires turning off and back on. Initially, staff had to request a ladder to go up into the ceiling and unplug the node. By 2003, RU's Physical Plant staff had run electrical piping to a toggle switch on the wall, which had made the process much easier.
Joseph Davis
Reference/Instruction Librarian
Roosevelt University Schaumburg Campus Library
Schaumburg, IL

Figure 5–2 is excerpted from the Microsoft troubleshooting Web site. The page shows some common problems you might encounter connecting a wireless network, and some possible causes and solutions. It covers both network-based problems (with access points or channel configuration) and end-user problems related to computer setup.

Figure 5–2   Excerpt from "How to Troubleshoot Wireless Connection Problems" (Microsoft.com 2005).

United States   Microsoft.com Home | Site Map

Microsoft   Excerpt from http://support.microsoft.com/default.aspx?scid=kb;en-us;831770   Search Microsoft.com for: [Go]

Help and Support   Note: This image has been edited for size.
Help and Support Home | Select a Product | Search Knowledge Base   The full article is available at the URL above.

MSBBN: How to troubleshoot wireless connection problems   Article ID   : 831770
Last Review : February 8, 2005
Revision   : 2.0

Verify that the wireless network is available

Note If your wireless adaptor is an integrated device such as a built-in wireless networking adaptor in a portable computer, you may have to turn it on by using a switch or a keyboard shortcut. See your computer's documentation, Help files, or technical support department.

1. Click **Start**, point to **Connect to**, and then click **Show all connections**.
2. Right-click your wireless adaptor, and then click **View Available Wireless Networks**.
3. Under **Available wireless networks**, verify that your network name (SSID) is displayed.
   - If your wireless network name is displayed, but you cannot connect to the network, one or more of the following conditions may be true:
     - The wireless security (WEP) settings on your computer may not be configured with the same information as your network. See the "Check the wireless security settings" section.
     - There may be a problem with the configuration of your wireless access point.
     - There may be a conflict between your wireless adaptor and another network adaptor on your computer. See the "Look for conflicts with other network adaptors" section.
   - If your wireless network name is not displayed, one or more of the following conditions may be true:
     - The wireless network may not be present.   **Reminder: The SSID is the unique name of**
     - The wireless access point may be configured not to broadcast the SSID.   **your wireless network, e.g. *acme_library*.**
     - There may be a problem with your wireless access point. Check all connections, and make sure that the power is on.
     - There may be a conflict between your wireless adaptor and another network adaptor on your computer. See the "Look for conflicts with other network adaptors" section.

Configure the wireless network settings on the base station

Change the wireless channel number

Note Channels 1, 6, and 11 are the preferred channels for wireless networking.   **See info on channels in Chapter 3 under Site Survey.**

Reduce the data rate

When you reduce the data rate on your wireless network, it may create a more stable connection. To do this, follow these steps:

1. Start the Microsoft Broadband Network Utility.
2. On the **Tools** menu, click **Management Tool**.
3. When you are prompted to, log on to the base station. To do this, use the password that you created when you ran the Setup Wizard the first time. If you did not run the Setup Wizard, use the following default base station password:

   admin

4. Click **Wireless**.   **This may be a step in your 'tweaking' process after**
5. Click to select the **Enable wireless access** check box if it is not already selected.   **you first install your WLAN.**
6. Change the data rate that your wireless network uses to a lower number. If your data setting is **auto**, first try **5**, and then try lower settings if you have to.
7. Click **Apply**.

If these steps resolve the problem, you may have a wireless signal quality or signal interference problem. Signals that are transmitted between the base station and a wireless adaptor can be affected by interference from other wireless devices—including 2.4 gigahertz (GHz) cordless phones, microwave ovens, and neighboring wireless networks. Move the other devices farther from your wireless networking hardware as needed, and do not use them while you are using the network.

Configure the wireless network settings on your computer

4. Examine the TCP/IP settings for your wireless network adaptor. To do this, follow these steps:
   a. In **Network Connections**, right-click your wireless adaptor, click **Properties**, and then click the **General** tab.
   b. Double-click **Internet Protocol (TCP/IP)** under **This connection uses the following items**.
   c. Use one of the following methods:
      - If your wireless network is configured to assign the IP address and DNS server information automatically (most common), click both of the following options:
        - **Obtain an IP address automatically**   **This is a very common error by patrons. This should be**
        - **Obtain DNS server address automatically**   **one of the first things you have them verify.**
      - If your wireless network is configured to use specific IP address and DNS server information (advanced), configure that information about this tab.

      Note This kind of configuration would have been set up by the person who configured your wireless network. Most wireless networks are automatically configured.

Manage Your Profile | Contact Us

©2006 Microsoft Corporation. All rights reserved. Terms of Use | Trademarks | Privacy Statement

> *Authors' Note—This is perhaps the most amusing troubleshooting issue we came across, though if you think about it, it's something to consider in your planning!*
> We have our access point in the back room and we had to train staff not to put things in front of it. Not because blocking the access point was a problem, but because they would bump it and it would get disconnected.
> Nicole Weber
> Assistant Director
> Clive Public Library
> Clive, IA

TROUBLESHOOTING FOR USERS

The question you will most frequently asked by patrons will be how to make their laptops connect to your wireless network. Before discussing some tips for creating effective troubleshooting documents that can help patrons help themselves, let's look at some common errors they may encounter.

Figure 5–3   Screenshot from Patron's Laptop with Incorrect IP.

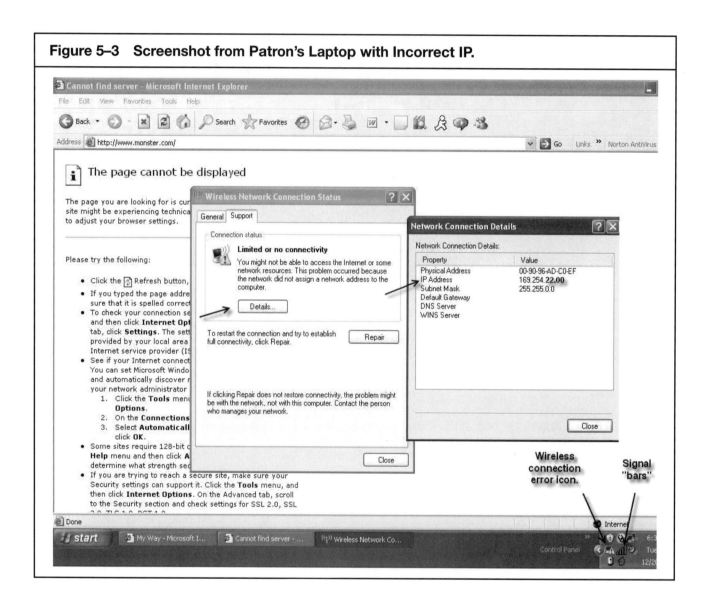

Figure 5–3 is a screenshot from a patron's laptop. The browser is not connecting, giving a "Cannot find server" DNS error. When we pull up the Wireless Connection Status, by double-clicking on the little wireless problem icon (an exclamation point in a yellow triangle next to the PC graphic), it says, "Limited or no connectivity." Looking in the system tray in the bottom right corner, we see that there are several "bars" of signal, so the signal strength is not the issue. When we click on "Details" an IP address and subnet are listed. However, that IP address (which we've changed for this demonstration) is not in the IP range of our network.

Therefore the patron's laptop is "holding on to" another IP, possibly because his laptop is set to manually set an IP in order to connect to an employer or university system. He first needs to update his wireless settings to accept DHCP (dynamic host configuration protocol) which will automatically assign an IP address. Here are the steps to do this:

- Go into your Network Connections (in XP, Start—Control Panel—Network Connections)
- Find the wireless connection from the list, and right-click on it. Choose "Properties" from the menu.
- Find the TCP/IP settings. It may say "Ethernet . . . TCP/IP." Click on it, then choose "Properties."
- In the XXX tab, click the radial buttons next to both "Obtain an IP address automatically" and "Obtain DNS server address automatically" (See Figure 5–4).

Figure 5–4   Enabling DHCP in Windows XP.

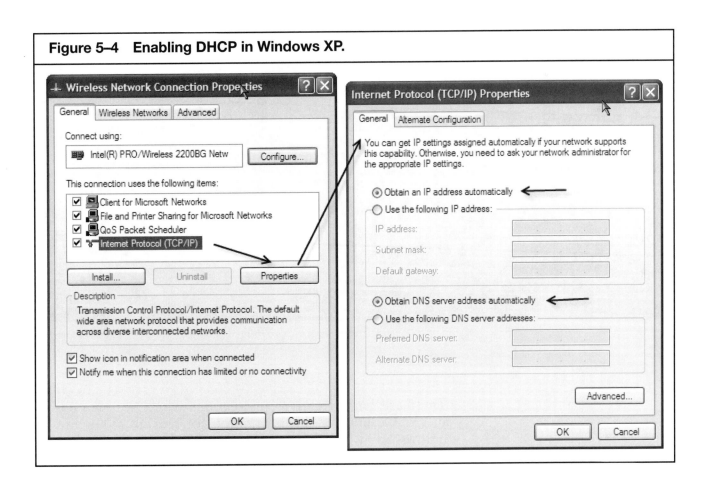

After this is set, you need to release and renew your IP settings. To do this:

- Go to a command prompt (Start—Accessories—Command Prompt).
- At the prompt, type "ipconfig /release," then Enter. At the next prompt, type "ipconfig /renew" in order to get a "good" IP address (see Figure 5–5).

Figure 5–5   Releasing and Renewing IP.

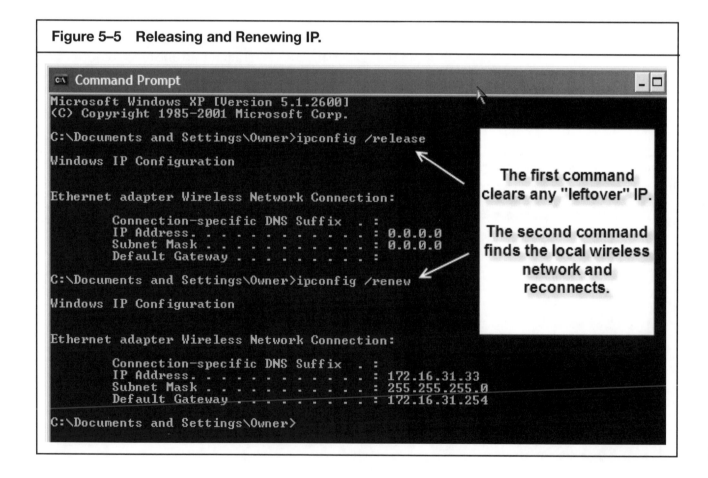

These are fairly standard instructions which you can add to your troubleshooting documentation for patrons. They are for Windows XP, which is the PC-based OS that generally has the easiest time connecting. The basic concepts will work for older versions of Windows although the setting may be in slightly different locations. See some of the troubleshooting examples in Source C for more ideas.

> *Authors' Note—SJCPL has help sheets they've created for their patrons which you can find in Source C. We heard this comment from many library staff:*
> We have found over the years that some of the older Windows laptops and older Windows operating systems have great difficulty connecting. There are no easy answers. You can work for twenty to thirty minutes helping someone get connected only to have the connection drop after a few minutes.
> Sue Hostetler
> Assistant Manager—Information Technology Services
> St. Joseph County Public Library
> South Bend, IN

Anyone who works with the public, especially librarians, knows that people have varied experience with technology which, in a public service environment like a library, must be accounted for and accommodated. As we've surveyed libraries with wireless networks and spoken to staff, we've found that the help offered to patrons about using a wireless hotspot covers a wide spectrum, from no assistance (and sometimes barely any mention of the service being available) to full hands-on assistance with configuring patron laptops by library staff. There is no right or wrong in this situation, simply differences in staffing, policy, and WLAN setup.

> Some staff were concerned we'd need to help people with access and we wouldn't know enough about the variety of wireless cards and associated laptops out there. Turned out nearly no one needs our help. We do require they type in an access code. We do this so they have to come to the desk and get our bookmark with the code that has info about security/liability/etc.
> Occasionally (twice per week) someone can't connect and we try to help in a limited way. Often we can't help, and that's just how it is. Sometimes it's a simple thing we can do, such as point out where their wireless icon is on the task bar. We don't touch their computers or do intensive problem solving.
> Liz Amundson
> Reference
> Madison Public Library
> Madison, WI

CREATING USEFUL FAQS

A common occurrence:

User complains he can't connect to wireless. Library staff looks at the computer and finds system invariably full of spyware. Patron is left shaking head wondering what in the world spyware is, and who put it on their computer.

"Last time I was here it worked! You must have put it on my computer!"

Warning patrons that they need to take steps to protect themselves against the viral and spyware, wild west world of the Internet is sometimes futile because the concepts are often beyond the users.

Users do NOT read agreements or policies, even if forced to, and even if they do, they often don't understand in the slightest the security concepts.

. . . Diagnosis pretty much involves making sure the patron receives an IP address (start-> run-> cmd-> ipconfig /all), making sure the wireless card is activated (more common than you'd think . . .) and ctrl-alt-del to see what is running. Spyware and Virii will more often than not make it impossible to have a successful connection attempt. We also find that the gateway just freezes up and requires reboot.

Hilariously enough, as I composed this e-mail . . . I was called out to address the exact situation.

Michael McEvoy
Electronic Services Support
Northville District Library
Northville, MI

As Michael McEvoy states in his user experience, "Users do NOT read agreements or policies." This may be true. However, you can "nip in the bud" many of the more common questions, complaints, and troubleshooting options with good documentation.

Once you've determined your security setup, start thinking like patrons, especially new users of your library. What information do they need to know in order to use your network safely? What information can you provide them with that would make them aware of any security or privacy issues they might encounter? Construct an FAQ (Frequently Asked Questions document) which you will post on your web site, at your service desks, and conspicuously in prime "hot spot" areas in your library. You know what signage works with your patrons (and what doesn't). This signage also provides a great marketing and public relations opportunity.

Are you requiring that patrons read and agree to user agreements before using your wireless service? As mentioned before, there are options for having a "click-through" page pop up where users have to read (or at least say that they have read) your policies and procedures before proceeding to the Internet at large or any of your networked resources. We would not recommend putting your FAQ solely on the click-through or splash page, as it is rarely actually read and the FAQ includes information they really should know.

> We don't expect our staff to be able to troubleshoot laptop problems because there are so many different operating systems and wireless cards out there that it would be difficult to keep everyone trained. Also, for liability purposes, we do not allow our staff to touch anybody's laptop. However, we do have a general troubleshooting handout for patrons in case there are problems. We also provide links to help pages from some of the major manufacturers on our Web page —www.wtamu.edu/library/circulation/wireless.shtml. (See #7 of the FAQ.)
> Beth Vizzini
> Reserves Supervisor
> Cornette Library
> West Texas A&M University

An effective FAQ lets patrons know how to connect to your wireless network and makes them aware of connectivity and security issues, including where "hot spots" are in the building, where they might encounter interference, whether they need a key to access your network, and so forth. In a similar fashion, simple troubleshooting guidelines for patrons give them basic instructions for troubleshooting their own equipment, should they have problems connecting. These troubleshooting guidelines should be created and made available in several ways, including:

- via your **Web site,**
- in **handouts** at all service desks, and
- on **signage** in popular "hot spot" areas in your library.

There is nothing secret or proprietary about simple troubleshooting "tips," so be sure to make them readily available to your patrons using wireless. Some libraries separate these guides from the general FAQ about connecting to the WLAN and/or wireless policy. If it's printed, this saves paper since you only hand it out to the minority experiencing problems. If online, you simply provide a link back and forth from your general FAQ and your troubleshooting guide (and also your computer/wireless use policy).

An excellent example of a troubleshooting guide can be found at the West Texas A&M University Web site. Figure 5–6 shows the guide, with our comments about its strengths. It is not the most comprehensive we found (see Source C for additional examples), but it covers the highlights and has some innovative items, like the maps for both access coverage and electrical outlets. It also gives tips for more than one kind of operating system, something many libraries forget.

Figure 5–6   Wireless Troubleshooting Guide. West Texas A&M University.

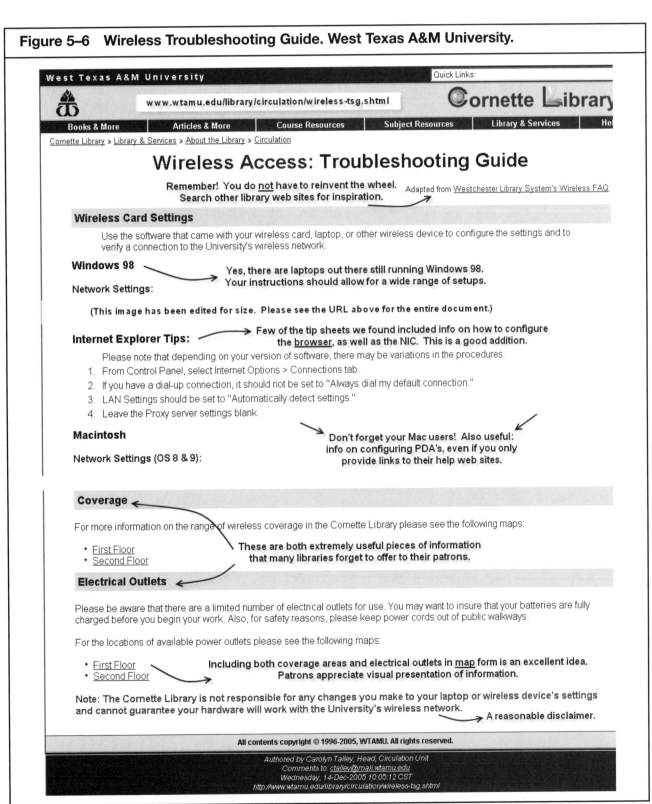

As mentioned in Beth Vizzini's experience at West Texas A&M, many libraries do not allow staff to touch patron's laptops. This is a policy issue. However, if you choose not to offer much staff assistance, your troubleshooting should include not only items we've mentioned above, but also a list of where they can get assistance (through university tech support, at any library service desk, by calling a support number for your ISP, etc.) and what hours this assistance is available.

> Patrons generally have an easy time connecting to the wireless if they know anything at all about their computers. We also offer a sheet of directions for them to use. I wouldn't say we have had a lot of problems with patrons accessing the network because they are usually a different clientele than those who come in to use our computers–generally the wireless users are business people or are students.
>
> My only recommendation would be adding more access points to increase the range of access for the laptops.
>
> Nicole Weber
> Assistant Director
> Clive Public Library
> Clive, IA

Figure 5–7   Standard Troubleshooting Help Sheet.

General Information for All Systems: Excerpt from West Des Moines (IA) Public Library FAQ
www.wdmlibrary.org/asp/faq/faq_details.asp?faqid=237&deptid=7

The following instructions are general guidelines and require that you understand how
to configure and restore the settings of your own computer. It is best to try to use the
wireless Internet service before attempting any of the changes suggested on this
page.

Wireless Card Settings:

Use the software that came with your wireless card or computer to configure the
settings and to verify a connection to the library's wireless network.

- SSID (network name) = **wdm_public**
- WEP = disable WEP encryption
- Mode or Network Type = Infrastructure mode or Access Point

Reminder: The SSID is the unique name
of your library wireless network. In this
case, Louise's library chooses to publish
the SSID of their openly-accessible WLAN,
so patrons know what they're looking for.

Windows 98 Network Settings:

1. From Control Panel, select Network.
2. Locate the TCP/IP protocol and go to Properties. If more than one TCP/IP
   protocol is listed, look for the one associated with your wireless adapter.
3. For IP Address, select "Obtain an IP address automatically."
4. For DNS, select "Disable DNS."
5. For WINS, select "Use DHCP for WINS Resolution."

Windows NT Network Settings:

1. From Control Panel, select Network.
2. Locate the TCP/IP protocol and get into its Properties. If more than one TCP/IP
   protocol is listed, look for the one associated with your wireless adapter.
3. Select "Obtain an IP address from a DHCP server" and click OK.

Windows 2000 & XP Network Settings:

1. From Control Panel, select "Network and Dial-up Connections" or "Network and
   Internet Connections > Internet Connections."
2. Right-click on "Wireless Network Connection" or "Local Area Connection" and
   click Properties.
3. Locate the Internet TCP/IP protocol and get into its Properties. If more than one
   TCP/IP protocol is listed, look for the one associated with your wireless adapter.
4. Select "Obtain an IP address automatically" and "Obtain DNS server address
   automatically."
5. For XP, right click on "Wireless Network Connection" and select "View available
   Wireless Networks." Select the name of the wireless network available at the
   library and click "Connect." The network/SSID name is **wdm_public**.

Page also includes Internet Explorer and Macintosh configuration tips.
See URL above for full document.

©Copyright 2006 City of West Des Moines | P. O. Box 65320 West Des Moines, Iowa 50265-0320 |Contact Directory | Privacy Policy

To review, your Frequently Asked Questions and Troubleshooting Tips (which may be a single document, or may be handed out separately as needed) may include any or all of the following items

- **Where the access is available:** a description, or better yet a map, of coverage areas in your building(s). You may also put this in your troubleshooting tips, if that's a separate document, to show them where the signal is "strongest." Why not? There's no harm in repeating useful information.
- **If you require an encryption key be entered, instructions on obtaining the key.** Something like "Show your library card to a library staff member to be given the WEP code" or "Go to this URL and sign in with your university login to obtain a WEP code."
- **Some helpful links to troubleshooting sites from Microsoft, Apple, Palm, etc.** You can't answer all their questions about their laptop's configuration.
- **What support is offered,** if any, and how to access it (online, phone, service desks). Also when it is available. If you have an ISP providing your wireless access, it may have a twenty-four hour support line. Some universities also have this service for students and faculty.
- **If support is not offered, a reiteration of this policy.** Something simple along the lines of "Library staff are not able to configure patron laptops. However, we offer these helpful tips you may want to try if you have difficulty connecting to our network."
- **A simple set of troubleshooting tips.** Figure 5–7 is an example of troubleshooting steps you can offer a patron. This particular set of steps is for a fully open WLAN.

As the many librarians' experiences we've shared with you attest, you will never satisfy all of your patrons' support needs, but if you do your best to provide options, you'll find them generally more satisfied with your library's service, even if they ultimately cannot connect their equipment. And when they can connect, you've added a valuable service to your library!

SOURCE A: GLOSSARY OF TERMS

3G

3G, which stands for the third generation of wireless communication technology, refers to pending improvements in wireless data and voice communications and includes a number of proposed standards. The third generation focuses on bringing high-speed connections and increasing reliability. 3G is also the name for a WWAN (Wireless Wide Area Network) that allows for high-speed data transmission and greater voice capacity for mobile users.

802.11

A group of wireless specifications developed by a group at the Institute of Electrical and Electronics Engineers (IEEE) in 1998. These specifications are used to manage packet traffic over a network and ensure that packets do not collide, which could result in loss of data, while traveling from device to device. As the standard 802.11 has evolved, a series of sub-standards designated by alphabetic characters has emerged:

- 802.11a—Operating in the five-GHz frequency range with a maximum fifty-four Mpbs signaling rate, this frequency band isn't as crowded as the 2.4-GHz frequency because it offers significantly more radio channels than the 802.11b and is used by fewer applications. It has a shorter range than 802.11g, is actually newer than 802.11b, and isn't compatible with 802.11b.
- 802.11b—Also known as "Wi-Fi," devices using this standard operate in the 2.4-GHz band and with rates of up to eleven Mbps. A very commonly used frequency; microwave ovens, cordless phones, medical and scientific equipment, and Bluetooth devices all work within the 2.4-GHz ISM band. Like other 802.11 standards, 802.11b uses the Ethernet protocol and CSMA/CA (carrier sense multiple access with collision avoidance) for path sharing.
- 802.11e—Ratified by the IEEE in September 2005, the 802.11e quality-of-service specification is designed to guarantee the quality of voice and video traffic. This will quickly become useful as companies develop products to use Voice Over IP (VoIP) for wireless connections.
- 802.11f—Ratified in 2003, the Inter-Access Point Protocol is a recommendation that describes an optional ex-

tension to IEEE 802.11 facilitating multi-vendor access point interoperability. It is used for roaming solutions.

- **802.11g**—This standard is similar to 802.11b but supports signaling rates of up to fifty-fouy Mbps. It also operates in the heavily used 2.4-GHz ISM band but uses a different radio technology to boost overall throughput, and is compatible with 802.11b, but not 802.11a.
- **802.11i**—Sometimes called Wi-Fi Protected Access 2 (see WPA 2), 802.11i was ratified in June 2004. WPA 2 supports the 128-bit-and-above Advanced Encryption Standard, along with 802.1x authentication and key management features.
- **802.11k**—Predicted for ratification in mid-2006, the 802.11k Radio Resource Management standard will provide measurement information for access points and switches to make wireless LANs run more efficiently.
- **802.11n**—The Standard for Enhancements for Higher Throughput is designed to raise effective WLAN throughput to more than one hundred Mbps. Final ratification is expected in 2006.
- **802.11r**—Expected to be ratified by late 2006, the 802.11r Fast Roaming standard will address maintaining connectivity as a user moves from one access point to another. This is especially important in applications such as voice-over-WLAN.
- **802.11s**—This standard will deal with mesh networking, and is predicted to be ratified in mid-2008.

802.15

802.15 is a communications specification that was approved in early 2002 by the Institute of Electrical and Electronics Engineers Standards Association (IEEE-SA) for wireless personal area networks (WPANs). The initial version, 802.15.1, was adapted from the Bluetooth specification (see **Bluetooth**). An explanation of the individual 802.15 standards comes from "The Free Dictionary."

- **802.15.1**—Bluetooth—Working with the Bluetooth SIG, the IEEE standardized the lower layers of the Bluetooth specification. Approved by the IEEE in 2002, 802.15.1 is fully compatible with Bluetooth 1.1.
- **802.15.2**—WPAN and WLAN Co-existence—Explores methods for interoperability between WPANs and WLANs as well as the ability to transmit in the same geographic area without interference.

- 802.15.3—High Rate Systems—A WPAN standard for data rates greater than twenty Mbps.
- 802.15.3a—Very High Data Rate-UWB-A variety of proposals have been reviewed for the ultrawideband (UWB) standard. Two competing methods are multiband ODFM, which divides the spectrum into 528 MHz bands, and a pseudo-random number sequence like CDMA (see UWB).
- 802.15.4—Low Data Rate Systems—A WPAN standard for devices that can run on batteries for months and years. Nodes can be configured as reduced function devices (RFDs), which communicate with full function devices (FFDs) only. FFDs can communicate with RFDs or FFDs.

802.16
See **WiMax**.

Ad hoc topology (or ad hoc mode)
Also known as independent basic service set (IBSS), this is wireless network framework in which devices can communicate directly with one another without using an access point (AP) or a connection to a regular network. Different from the more common infrastructure network in which all devices communicate through an AP, libraries would rarely use this mode, except in some staff and inventory projects, or for connecting wireless peripherals as needed.

AMPS (Analog Mobile Phone System)
A term used for analog technologies, the first generation of wireless technologies.

Antenna
A device that facilitates the transmission and reception of radio signals. May be omnidirectional (transmitting in many directions) or directional (transmitting in one or few directions—to focus signal in elongated or odd area configurations).

AP (Access point)
For wireless Ethernet, a transceiver device known as an access point is connected to the wired network using a standard Cat5 cable. Most APs are about the size of a book and include a stubby antenna. The access point, or transceiver, is itself either a switch or a hub that, at a minimum, reinforces the signal and transmits the data between users on the wireless network, one or many wireless devices, and the Internet backbone (usually wired Ethernet).

Attenuation

The dissipation of the strength of a signal as it is transmitted.

Authentication

Authentication, fundamentally, is the process of determining whether someone or something is, in fact, who or what it is declared to be. In computer security, it is verification of the identity of a user or the user's eligibility to access an object. Because password protection, long the basic standard authentication method, is open to much human error and hacking, more secure networks require a more stringent authentication process. The most common form of authentication is user name and password, although this also provides the lowest level of security. VPNs use digital certificates and digital signatures to more accurately identify the user. Authentication is distinct from authorization, which is the process of giving individuals access to system objects based on their identity, though these functions may appear seamless to the user.

Bands (radio bands)

A band is a small section of the electromagnetic spectrum of radio communication frequencies, in which channels are usually used or set aside for the same purpose.

Bandwidth

In wired networks, the size of a network "pipe" or medium for communication. In wireless, it describes the transmission capacity in terms of a range of frequencies. Used to indicate how much, or how fast, data can be transmitted across a telecommunications line or network connection in a period of time, usually one second; sometimes used synonymously with data transfer rate, throughput, and line speed.

BlackBerry ™

Two-way wireless device, made by Waterloo, Ontario–based company Research in Motion, that allows users to check e-mail and voicemail (translated into text), as well as page other users via a wireless network service. Also known as a RIM device, it has a miniature qwerty keyboard for users to type their messages. It uses the SMS protocol (see **SMS**). BlackBerry users must subscribe to a wireless service that allows for data transmission.

Bluetooth

A low-cost, short-range wireless specification (in the 2.4 GHz range) that allows for radio connections between devices within a ten meter (about thirty foot) range of each other, including laptops, APs, phones, and printers. Can be used to create ad hoc wireless networks for printing, PDA downloads, etc. The Bluetooth specification was ratified into the IEEE 802.15 communications specification, which is fully compatible with Bluetooth 1.1. The name comes from tenth-century Danish King Harald Blåtand (Bluetooth), who unified Denmark and Norway.

bps

Bits-per-second—a common unit of measure for data transfer rates used in the U.S., as in Kbps (kilobits per second) and Mbps (megabits per second).

Broadband

A communications channel of high bandwidth, capable of transmitting a relatively large amount of data over a given period of time.

BSS (Basic Service Set)

Communicating stations on a wireless LAN. See also infrastructure basic service set.

BSS (Infrastructure Basic Service Set)

A type of IEEE 802.11 network comprised of both stations and access points (APs) that are used for all communication within the BSS, even if the stations reside within the same area.. Communication from station to station takes place through the APs.

Captive portal

A method of "capturing" your Web traffic and forcing users to "click-through" a user agreement, authenticate, and/or pay for service before they can access the Internet at large. If you've ever used the Internet service at a Starbucks or a Borders store, you have encountered a captive portal. A captive portal forces an HTTP client on a network to see a special Web page before being released out to the Internet. This is done by intercepting all HTTP traffic, regardless of address, until the user is allowed to exit the portal. Usually captive portals are used for authentication.

Cat5 (or Category 5) cable

Unshielded twisted pair cable used in many local area network (LAN) installations. Slowly being replaced by Cat5e (designed for Gigabit Ethernet cabling) and Cat6 cable.

CDMA (Code Division Multiple Access)

According to Wikipedia.org, CMDA is "a method of multiple access that does not divide up the channel by time (as in TDMA), or frequency (as in FDMA), but instead encodes data with a certain code associated with a channel and uses the constructive interference properties of the signal medium to perform the multiplexing. CDMA also refers to digital cellular telephony systems that make use of this multiple access scheme." Phones and other devices that have a cellular radio built in use GSM/CDMA.

CDPD (Cellular Digital Packet Data)

Cellular network technology that uses unused bandwidth normally used by AMPS mobile phones between eight and nine hundred MHz to transfer data. Speeds up to 19.2 Kbps are possible. CDPD is largely unused now, as faster standards such as GPRS have become dominant.

Channels

A specific band of the radio frequency spectrum used for radio transmissions.

Circuit-switched network

A type of network, such as the regular voice telephone network in which the communication circuit (path) for the call is set up and dedicated to the participants in that call. For the duration of the connection, in contrast packet-switched network, all resources on that circuit are unavailable for other users.

Communications medium

The electronic component used to connect two network devices to facilitate network data transmission it usually refers to some type of data cabling (coaxial, thinnet, or twisted pair) or telecommunications line: In RF (radio frequency) wireless networking, it refers to radio energy.

CSMA/CA and CSMA/CD

Wired Ethernet uses a control medium known as "carrier sense multiple access with collision detection" while wireless LANs use "carrier sense multiple access with collision avoidance." In CSMA/CD (wired), the devices at the data link and physical layers "listen" to what is being transmitted on the wire in order to avoid transmitting at the same time as another device on the network. When two devices transmit at exactly the same time, it results in what is known as a collision, and both devices must retransmit the data. With wireless (CSMA/CA), once a node begins transmitting, it cannot detect whether or not another node is also in the process of transmitting, so wireless LANs rely upon the receiver to acknowledge receipt of the transmission.

Data transfer rate, or data rate

The number of bits that can be transferred across a network connection in one second; commonly used as a synonym for throughput or bandwidth.

DHCP (Dynamic Host Configuration Protocol)

This software that automatically assigns temporary IP addresses to client stations logging onto an IP network eliminates having to manually assign permanent "static" IP addresses. DHCP software runs in servers and routers.

DSSS (Direct Sequence Spread Spectrum)

In this transmission method, the stream of information to be transmitted is divided into small pieces, each of which is allocated to a frequency channel across the spectrum. A data signal at the point of transmission is combined with a code that divides the data according to a spreading ratio. The code helps the signal resist interference and also enables the original data to be recovered if data bits are damaged during transmission. While it is more expensive and uses more power than frequency hopping spread spectrum, it is also more reliable.

EAP (Extensible Authentication Protocol)

According to Wikipedia.org, "EAP (pronounced 'eep') is a universal authentication mechanism, frequently used in wireless networks and point-to-point connections. Although the EAP protocol is not limited to wireless LAN networks and can be used for wired LAN authentication, it is most often used in wireless LAN networks. Recently, the WPA and WPA2 standard has officially adopted five EAP types as its official authentication mechanisms."

Electromagnetic spectrum

The electromagnetic spectrum describes the range of all forms of radiation from sound to radio waves, through visible light, to harmful x-rays and gamma rays. Different sections of the spectrum are called bands, containing a range of frequencies within the spectrum that can be classified, such as visible light or infrared radiation.

Encryption

In cryptography, encryption is the process of obscuring information to make it unreadable without special knowledge. In the mid-1970s, strong encryption moved from use largely by government agencies into the public domain, and is now employed in protecting widely used systems, such as Internet e-commerce, mobile telephone networks, and bank automatic teller machines (see **Key**).

ESS (Extended Service Set)

A wireless network topology using multiple access points to connect devices to the wireless LAN. In an ESS infrastructure, each access point acts as a sort of relay station, determining if data should be sent to stations within its BSS, or passed on to other access points or perhaps to the wired network. Users can move throughout the entire WLAN area, from one access point to another, without losing their connections.

Ethernet

A standard computer networking communication format, it is the most common network format in use today.

FCC (Federal Communications Commission)

The U.S. government agency responsible for regulating communications industries.

FDMA (Frequency Division Multiple Access

FDMA is the division of the wireless cellular telephone band into 30 channels, each of which can carry a voice conversation or, with digital service, digital data. Used in the analog Advanced Mobile Phone Service (AMPS) phone system. With FDMA, each channel can be assigned to only one user at a time. D-AMPS (Digital-Advanced Mobile Phone Service) also uses FDMA but adds time division multiple access (TDMA) to get three channels for each FDMA channel instead of just one.

FHSS (Frequency Hopping Spread Spectrum)

A transmission method by which a carrier "hops" packets of information (voice or data) over different frequencies, according to an algorithm.

Frequency

The measurement of the number of times that a repeated event occurs in a fixed period of time, measured in Hertz (Hz). With radio, the frequency is the number of cycles of the repetitive wave-form per second.

Gain

A measure of the amount of focus an antenna uses transmitting/receiving a radio signal.

GPRS (General Packet Radio Service)

A mobile data service available to users of GSM standard mobile phones (see **GSM**). It is often described as "2.5G," meaning, a technology between the second (2G) and third (3G) generations of mobile telephony. It provides moderate speed data transfer by using unused TDMA channels in the GSM network.

GSM (Global System for Mobile Communications)

The definition from Wikipedia.org states, "The Global System for Mobile Communications (GSM) is the most popular standard for mobile phones in the world. GSM service is used by over 1.5 billion people across more than 210 countries and territories. The ubiquity of the GSM standard makes international roaming very common between mobile phone operators, enabling subscribers to use their phones in many parts of the world. GSM differs significantly from its predecessors in that both signaling and speech channels are digital, which means that it is considered a second generation (2G) mobile phone system. This fact has also meant that data communication was built into the system from very early on. GSM is an open standard which is currently developed by the 3GPP."

Hertz

A unit of measure of electromagnetic frequency named after Heinrich Hertz. One hertz corresponds to one cycle per second (or one wave per second).

HomeRF

A home networking standard developed by Proxim Inc. that combines the 802.11b and Digital Enhanced Cordless Telecommunication (DECT) portable phone standards into a single system. HomeRF uses a frequency hopping technique to deliver speeds of up to 1.6 Mbps over distances of up to 150 feet (2.4 GHz). This short range makes it useful primarily for the residential market, where it vies with 802.11b "Wi-Fi." It includes technologies that reduce interference from home appliances.

Hot spot

A place, such as a hotel, restaurant, or airport, that offers Wi-Fi access, either free or for a fee.

IBSS

Independent basic service set (see **Ad hoc mode**).

IEEE

The Institute of Electrical and Electronics Engineers is a nonprofit, technical professional association of more than 360,000 individual members in approximately 175 countries that is an authority in technical areas such as computer engineering and telecommunications. It developed the 802.11 specifications.

I-Mode

A wildly popular service in Japan for transferring packet-based data to handheld devices, and the first smart phone for Web browsing. I-Mode is based on a compact version of HTML and does not use WAP (Wireless Application Protocol), setting it apart from other widely used transmission methods. It uses a simplified version of HTML, Compact Wireless Markup Language (CWML) instead of WAP's Wireless Markup Language (WML). I-Mode's creator, NTT DoCoMo of Tokyo, recently bought large shares in AT&T and plans to move in on the U.S. market, though no date is set.

ISP (Internet Service Provider)

A commercial or public agency providing connections to the Internet, either via network or dial-up services; services vary widely.

Key

In encryption, a key specifies the particular transformation of plaintext into ciphertext, or vice versa during decryption. Keys are also used in other cryptographic algorithms used for authentication.

LAN (Local Area Network)

A group of computers connected over a communications medium for the purpose of sharing access to centralized resources (files, printers, CD-ROM products).

Line of sight, or radio line of sight

The visible, unobstructed path required to enable radio communications between two distinct access points.

MAC

Every wireless 802.11 device has its own specific Media Access Control address hard-coded into it. This unique identifier can be used to provide security for wireless networks. When a network uses a MAC table, only the 802.11 radio devices that have had their MAC addresses added to that network's MAC table are able to get onto the network. This is a minimum level of security.

Mesh networking

A wireless mesh network is a collection of wireless devices maintaining RF connectivity to create a seamless path for data packets to travel. At least one wireless device (or node) is connected to a wired Internet backbone. In the wireless mesh environment, a network can be envisioned as a collection of access points, routers, or end-users (equipped with wireless receivers/transmitters) that are free to move arbitrarily but maintain a reliable communication that sends and receive messages. The 802.11s standard is planned to address mesh networking issues but has not yet been ratified.

MIMO

Multiple input multiple output refers to using multiple antennas in a Wi-Fi device to improve performance and throughput. The MIMO technology takes advantage of a characteristic called multipath, which occurs when a radio transmission starts out at point A and then reflects off or passes through surfaces or objects before arriving, via multiple paths, at point B. MIMO technology uses multiple antennas to send and receive two or more unique data streams over the same channel simultaneously in wireless devices, resulting in networks with long ranges and high throughputs. In some cases, wireless networks using MIMO technology are expected to reach over three hundred feet and still send and receive data at thirty Mbps. The technology is the basis of the forthcoming 802.11n standard.

Network protocol

A set of rules used for the process of packaging, starting, interrupting, and continuing network data communications.

NIC

A network interface card is an internal or external hardware card that translates radio signals sent via an access point to your computer. Recently, most laptops come with wireless NICs installed.

OEMs

Original equipment manufacturers.

Packet

A chunk of data that is sent over a network, whether it's the Internet or wireless network.

Packet-switched network

A type of network in which relatively small units of data called packets are routed through a network based on the destination address contained within each packet. Breaking communication down into packets allows the same data path to be shared among many users in the network. Most traffic over the Internet uses packet switching rather than circuit-switched networking. Voice calls using the Internet's packet-switched system are possible (VoIP, or Voice-over IP, uses this). Each end of the conversation is broken down into packets that are reassembled at the other end.

PCI cards

Designed by the Intel company, PCI (peripheral component interconnect) is an interconnection system between a microprocessor and attached devices in which expansion slots are spaced closely for high-speed operation. Using PCI, a computer can support both new PCI cards while continuing to support Industry Standard Architecture (ISA) expansion cards, an older standard. The standard describes two different card lengths.

PCMCIA card

A credit card-sized memory or I/O device that connects to a personal computer, usually a notebook or laptop computer. The PCMCIA 2.1 standard was published in 1993 with the goal that PC users can be assured of uniform attachments for any peripheral device that follows the standard. A PCMCIA card has a sixty-eight-pin connector that connects into a slot in the PC. There are three sizes (or "types") of PCMCIA cards, for different devices. The PCMCIA standard is most commonly applied to portable PCs but it can also be used on desktop computers.

PDA (Personal Digital Assistant)

Mobile, handheld devices such as the Palm series and Handspring Visors. Users can synchronize their PDAs with a PC or network; some models support wireless communication to retrieve and send e-mail and get information from the Web.

Phase-shift key

In direct sequence spread-spectrum transmissions, the phase-shift key is a pseudo-random code generated by the sender. That code is required by the receiver to interpret the signal.

Plenum

The area above a room where heating and air conditioning run. Any device placed in the plenum area must meet the UL2043 standard for fire safety.

Point-to-point

A network connection that is tied directly to two locations.

Radio frequency spectrum

The portion of the electromagnetic spectrum from approximately thirty KHz to one GHz.

Radio frequency communications

The use of radio energy to carry voice and data signals between two or more access points.

RF (Radio frequency)

Radio frequency, or RF, refers to that portion of the electromagnetic spectrum in which electromagnetic (radio) waves can be generated by alternating current fed to an antenna.

RF noise

Undesired radio signals that alter a radio communications signal causing extraneous sounds during transmission and/or reception.

RFID (Radio Frequency IDentification)

A wireless data collection technology that uses electronic tags for storing data. Like bar codes, they are used to identify items. Unlike bar codes, which must be brought close to the scanner for reading, RFID uses low-powered radio transmitters to read data stored in the transponder (tag) at distances ranging from one inch to one hundred feet.

Radio frequency devices

These devices use radio frequencies to transmit data.

Roaming

Movement of a mobile device from one wireless network location to another without interruption in service or loss in connectivity.

Site survey

Review done before installing or expanding a WLAN, with the purpose of exploring line of sight issues for access points, transmission distances for wireless transmitters, and sources of potential interference, including physical structures and other radio signals. It generally involves diagramming the network and the building, checking building plans, and testing the equipment and signal strength.

Smart phone

A combination of a mobile phone and a PDA, the smart phone allow users to converse as well as perform tasks, such as accessing the Internet wirelessly and storing contacts in databases. The increasingly popular smart phones have PDA-like screens.

SMS (Short Messaging Service)

A service through which users can send text-based messages from one device to another (see **BlackBerry**). The message—up to 160 characters—appears on the screen of the receiving device.

Spectrum analysis

This process of analyzing a portion of the radio spectrum to determine if another agency may be using it for communications can be part of a site survey.

Spectrum analyzer

This device, which searches a band of radio frequencies for the presence of radio signals, may be used in a site survey to determine if there are conflicting signals that might interfere with the network signal.

SSID (Service Set Identifier)

This sequence of characters that uniquely describes a wireless network allows all attached devices to identify themselves and connect to the correct network when more than one independent network is operating in nearby areas.

Subnet

A portion of a network that shares a common address component. On TCP/IP networks, subnets are defined as all devices whose IP addresses have the same prefix. For example, all devices with IP addresses that start with 100.100.100. would be part of the same subnet. Dividing a network into subnets is useful for both security and performance reasons. IP networks are divided using a subnet mask.

T–1 lines

A special type of telephone line called a data circuit that connects two distinct points to allow for intercommunication consisting of both voice and data signals.

TDMA (Time Division Multiple Access)

A technology for shared medium (radio/wireless) networks, TDMA allows several users to share the same frequency by dividing it into different time slots. The users transmit in rapid succession, one after the other, each using their own time slot. This allows multiple users to share the same radio frequency while using only the part of its bandwidth they require. This technology is used in GSM networks.

Throughput

Commonly used as a synonym to data transfer rate or bandwidth. In wireless applications it refers to the actual quantity of data that can be transmitted over a wireless link.

Time Hopping Spread Spectrum (THSS)

Using what is commonly referred to as Ultra-Wide Bandwidth (UWB), time hopping (sometimes referred to as "digital pulse" or "impulse radio") uses extremely short pulses, or "chirps," where the signal is switched on and off in rapid succession in a digital application. The pattern on the on/off states of the frequency can be used to transmit data digitally where the early arrival of a pulsed signal represents a "0" and a late arrival represents a "1."

TKIP (Temporal Key Integrity Protocol)

WPA corrected some deficiencies in the older Wired Equivalent Privacy (WEP) standard, including adding the TKIP security protocol. TKIP hashes the initialization vector (IV) values with the WPA key to form the identifying traffic key, addressing one of WEP's largest security weaknesses. WEP simply concatenated its key with the IV, which was less complex, and easier to break. With TKIP, every data packet is sent with its own unique encryption key, so is more secure.

UWB

Ultrawideband, also called digital pulse, is a wireless technology for transmitting digital data over a wide swath of the radio frequency spectrum with very low power. Because of the low power requirement, it can carry signals through doors and other obstacles that tend to reflect signals at more limited bandwidths and a higher power. It can carry large amounts of data and is used for ground-penetrating radar and radio locations systems.

VLAN

A logical, not physical, group of devices defined by software, Virtual LANs allow network administrators to resegment their networks without physically rearranging the devices or network connections. VLANs are configured through software rather than hardware, so they are very flexible. One of the biggest advantages of VLANs is that when a computer is physically moved to another location, it can stay on the same VLAN without any hardware reconfiguration.

VPN (Virtual Private Network)

A network that uses a public telecommunication infrastructure, such as the Internet, to provide remote offices or individual users with secure access to their organization's network. The goal of a VPN is to provide the organization with the same capabilities as a private network, but at a much lower cost. A VPN works by using the shared public infrastructure while maintaining privacy through security procedures and tunneling protocols. Data, encrypted at the sending end and decrypted at the receiving end, is sent through a "tunnel" that cannot be "entered" by data that is not properly encrypted. An additional level of security involves encrypting not only the data, but also the originating and receiving network addresses.

WAP (Wireless Application Protocol)

The Wireless Application Protocol is a set of specifications, developed by the WAP Forum, that lets developers using wireless markup language (WML) build networked applications designed for handheld wireless devices. WAP was designed to work within the constraints of these devices: a limited memory and CPU size, small, monochrome screens, low bandwidth and erratic connections. WAP works with multiple standards.

War chalking

Marking buildings or sidewalks with chalk to show others where it's possible to access an exposed company wireless network. These access points are typically found through war driving.

War driving

Typically refers to driving around with a wireless-enabled laptop and antenna to locate and exploit security-exposed WLANs.

Wave

The basic component of the electromagnetic spectrum, waves (such as radio waves) travel and transfer energy from one point to another. Waves are characterized by crests and troughs, either perpendicular or parallel to wave motion.

Wavelength

The distance in the line of advance of a wave from any one point on a wave to a corresponding point on the next wave.

WECA (Wireless Ethernet Compatibility Alliance)

See **Wi-Fi Alliance**.

WEP (Wired-Equivalent Privacy)

Wired-Equivalent Privacy protocol was specified in the IEEE 802.11 standard to provide a WLAN with a minimal level of security and privacy comparable to a typical wired LAN, using data encryption. It's now widely recognized as flawed because of an insufficient key length and other problems, and can be cracked in a short time with readily available tools. (see **WPA**).

Wi-Fi

Short for "wireless fidelity," the generic term for 802.11 technology. Originally describing the 802.11b standard, it is now used generically for the overall technology. Many airports, hotels, and other locations all known as hot spots offer public access to Wi-Fi networks so people can log onto the Internet and receive e-mails on the move.

Wi-Fi Alliance

A nonprofit association originally formed in 1999 under the name Wireless Ethernet Compatibility Alliance (WECA) to promote the growth of wireless LANs and to certify interoperability of WLAN products based on the IEEE 802.11 specification. The goal of the Wi-Fi Alliance's members is to enhance the user experience through product interoperability.

WiMax (Worldwide Interoperability for Microwave Access)

Popular name for broadband wireless networks based on the IEEE 802.16 wireless metropolitan-area network (WMAN) standard. IEEE 802.16 is working group number 16 of IEEE 802, specializing in point-to-multipoint broadband wireless access, and aims to improve on 802.11 (Wi-Fi) both in performance and coverage distances. WiMAX is designed to extend local Wi-Fi networks across greater distances, such as a campus or municipality, as well as to provide last mile connectivity to an ISP or other carrier many miles away. (See also **WMAN**)

Wireless bridge

A combination device that includes a network bridge and a radio transceiver, a bridge connects a local area network (LAN) to the radio link.

Wireless Ethernet

Wireless network connections that provide a throughput of five to seven Mbps, which is similar to the data throughput of standard, wired Ethernet networks.

Wireless LAN.

See **WLAN**.

Wireless NIC

Wireless network interface cards (see **NIC**). The vast majority of laptops sold in the last two years have wireless NIC factory-installed.

Wireless spectrum

A band of frequencies where wireless signals travel carry voice and data information.

WLAN

Wireless local area networks use radio waves instead of a cable to connect a user device, such as a laptop computer, to a network. They provide Ethernet-type connections over the air and operate under the 802.11 family of specifications developed by the IEEE. A wireless LAN can serve as a replacement for or extension to a wired LAN.

WMAN

The IEEE 802.16 wireless metropolitan-area network (WMAN) standard is popularly known as WiMax. It is used in many current initiatives by municipalities to provide free or low-cost wireless broadband access to residents over a broad geographic area, sometimes known as "municipal Wi-Fi." This can include putting access points, network bridges and other equipment on water towers, rooftops, etc. throughout a city center or in areas where broadband cabling has not yet been extensively laid. (See also **WiMax**.)

WML (Wireless Markup Language)

Wireless markup language is an Internet programming language used specifically for wireless product applications. Wireless application developers use WML to repurpose content for wireless devices. It delivers Internet content to small wireless devices such as browser-equipped cellular phones and handheld devices which typically have very small displays, slow CPUs, limited memory capacity, low bandwidth, and restricted user-input capabilities.

WPA (WiFi Protected Access)

Wi-Fi protected access is a data encryption specification for 802.11 wireless networks that replaces the weaker WEP. Created by the Wi-Fi Alliance before the 802.11i security standard was ratified by the IEEE, it improves on WEP by using dynamic keys, extensible authentication protocol (EAP) to secure network access, and an encryption method called temporal key integrity protocol (TKIP) to secure data transmissions.

WPA2

Wi-Fi protected access 2 is an enhanced version of WPA. It is the official 802.11i standard that was ratified by the IEEE in June 2004. It uses advanced encryption standard instead of TKIP (see **WPA**), which supports 128-bit, 192-bit and 256-bit keys. The WPA logo certifies that devices are compliant with a subset of the IEEE 802.11i protocol. WPA2 certifies full support for 802.11i.

WPA and WPA2 use a key hierarchy that generates new encryption keys each time a mobile device establishes itself with an access point. Like WEP, keys can still be entered manually (preshared keys); however, using a RADIUS authentication server provides automatic key generation and enterprise-wide authentication.

WPAN

A wireless personal area network is typically limited to short-range needs and is largely used to connect wireless devices in a single work area. A WPAN might be used to transfer data between a handheld device and a wireless-enabled desktop machine or printer. It could also be used to send data from wired PCs to a wireless-enabled printer in the room, which is also being used for printing by wireless devices (laptops, PDAs, smart phones). The receiving device (printer in this case) would need a wireless network card to connect. WPANs are sometimes used in homes to provide wireless connections for home security alarms, appliances, and entertainment systems. Typically, a WPAN uses some technology that permits communication within a short range of about ten meters, or thirty feet. The primary technology for WPAN is Bluetooth, from which specification the IEEE 802.15 standard was created, approved in early 2002.

WWAN

A wireless wide area network is a form of wireless network which differs from a WLAN (wireless local area network) in that it uses cellular network technologies such as GPRS, CDMA, GSM, or CDPD to transfer data. These cellular technologies are offered by cellular service providers for a monthly usage fee. They go beyond the basic cellular telephony and include data transfer, and since they're offered by nationwide cellular companies, can offer nationwide coverage and service. Some computers now have integrated WWAN capabilities, which means they have a cellular radio (GSM/CDMA) built in, allowing the user to send and receive data.

SOURCE B: READ MORE ABOUT IT

Below are works cited in the preceding chapters, as well as helpful resources used in researching this book. Keep in mind that this is a new technology, relatively speaking, so database searches for additional, more current articles and resources would be fruitful and recommended. Also see the next section, where helpful journals, Web sites, and e-mail lists are mentioned to help keep you current with new wireless issues and initiatives, in the library world and elsewhere.

WORKS CITED AND ADDITIONAL RESOURCES

"802.15 definition of 802.15 in computing dictionary—by the Free Online Dictionary, Thesaurus and Encyclopedia." The free dictionary.com Accessed: February 12, 2006. Available: reedictionary.com" http://computing-dictionary.thefreedictionary.com.

Allmang, Nancy. "Our Plan for a Wireless Loan Service." *Computers in Libraries* 23 no. 3 (March 2003): 20.

Allmang, Nancy. "Wireless Loan Service-Updating Our Story." Leter to the Editor (follow-up to 2003 article). *Computers in Libraries* 25 no. 7 (July/August 2005): 50–52.

Balas, Janet I. "Looking Ahead to New Technologies." *Computers in Libraries* 21 no. 3 (March 2001): 57.

Bertot et al. "Public Libraries and the Internet 2002: Internet Connectivity and Networked Services." *Information Use Management and Policy Institute.* (December 2002). Available (PDF): www.ii.fsu.edu/Projects/2002pli/2002.plinternet.study.pdf .

Boss, Richard W. "Mobile Computer Devices in Libraries." *ALA Technotes.* (November 2004). Available: www.ala.org/ala/pla/plapubs/technotes/mobilecomputer.htm.

Boss, Richard W. "Wireless LANs." *ALA Technotes.* (November 2004) Available: www.ala.org/ala/pla/plapubs/technotes/wirelesslans.htm.

Bradley, Tony and Becky Waring. "Complete Guide to Wi-Fi Security." *Jwire.com* (December 6, 2005). Available: www.jiwire.com/wi-fi-security-introduction-overview.htm.

Brain, Marshall. "How Wi-Fi Works." *Howstuffworks.com.* Accessed: October 15, 2005. Available: electronics.howstuffworks.com/wireless-network.htm/.

Breeding, Marshall. "A hard look at wireless networks." *Library Journal, Net Connect.* Summer 2002: 14–17.

Breeding, Marshall. "Implementing Wireless Networks Without Compromising Security." *Computers in Libraries* 25, no. 3 (2005): 34–36.

Breeding, Marshall. "Protecting Personal Information." *Computers in Libraries* 24 no. 4 (April 2004): 22–24.

Breeding, Marshall. "Wireless Networks Connect Libraries to a Mobile Society." *Computers In Libraries* 23, no. 9 (2004). Available: www.infotoday.com/cilmag/oct04/breeding.shtml.

Breeding, Marshall. "Wireless Networks in Libraries." ALA Library Technology Reports 41:5. September/October 2005.

"Captive Portal." *Wikipedia.org.* (December 29, 2005). Available: wikipedia.org/wiki/Captive_portal.

Chaplin, Kevin. *Wireless LANs vs. Wireless WANs.* Sierra Wireless White Paper 2130273. (November 18, 2002) Available: www.sierrawireless.com/news/docs/2130273_WWAN_v_WLAN.pdf.

Coe, Lewis. *Wireless Radio: A Brief History.* Jefferson, N.C.: McFarland, 1996.

DeGroff, Amy Begg. "Howard County Library WiFi Policy." *WebJunction.* (July 25, 2005) Available: www.webjunction.org/do/DisplayContent?id=11059.

Dornan, Andy. *The Essential Guide to Wireless Communications Applications : From Cellular Systems to Wi-Fi.* Upper Saddle River, NJ : Prentice Hall PTR, 2002.

Edelblute, Thomas. "Library Computer Without Wires." *InfoPeople Webcast* (March 1, 2005). Available: www.infopeople.org/training/webcasts/03–01–05/03–01–05_wifi.html.

Electronic Frontier Foundation (EFF). "Movie Legend Hedy Lamarr to be Given Special Award at EFF's Sixth Annual Pioneer Awards." Electronic Frontier Foundation Media Release. (March 12, 1997). Available www.eff.org/awards/pioneer/1997.php.

Embrey, Theresa A. Ross. "Today's PDAs Can Put an OPAC in the Palm of Your Hand." *Computers in Libraries* 22, no. 3 (March 2002): 14–20.

"Enterprise Wireless WAN Security." *HP.com documents.* Accessed: December 3, 2005 Available: http://h18004.www1.hp.com/products/wireless/wwan/WWAN-Security.pdf.

Federal Communications Commission (FCC). *Third Generation Wireless Systems.* Accessed: October 15, 2005 Available: www.fcc.gov/3G/.

Friedman, Matthew. "Protect Yourself Against Wi-Fi Bandwidth Vampires." *Networking Pipeline.* (August 9, 2005) Available: www.networkingpipeline.com/showArticle.jhtml?articleID=168200045.

Geier, Jim. "802.11 WEP: Concepts and Vulnerability." *Wi-Fi Planet.com.* (June 20, 2002). Available: www.wi-fiplanet.com/tutorials/article.php/1368661.

Geier, Jim. "Mobile IP Enables WLAN Roaming." *Wi-Fi Planet.com.* (May 13, 2003) Available: www.wi-fiplanet.com/tutorials/article.php/2205821.

Geier, Jim. "RF Site Survey Steps." *Wi-Fi Planet.com.* Accessed: May 10, 2002. Available: www.wi-fiplanet.com/tutorials/article.php/1116311.

Ginzburg, Barbara. "Goin' Mobile: Using a Wireless Network in the Library." *Computers in Libraries* 21, no. 3 (2001): 40–44.

Glas, Jack. "The Principles of Spread Spectrum Communication." Available: here http://cas.et.tudelft.nl/~glas/ssc/techn/techniques.html.

Beaumont, Chris. "Hedy Lamarr, George Antheil and the Secret Communications System Patent." Accessed: September 15, 2005. Available: www.ncafe.com/chris/pat2/index.html.

"Home RF 2.0" *PC Magazine*. (November 27, 2001) Available: www.pcmag.com/article2/0,1759,5198,00.asp

"How to troubleshoot wireless network connections in Windows XP." *Microsoft.com*. (June 22, 2004) Available: http://support.microsoft.com/ default.aspx?scid=kb;en-us;Q313242.

Institute for Electrical and Electronics Engineers (IEEE). *802.11 Working Group*. Accessed: September 7, 2005. Available. http://grouper.ieee.org/ groups/802/11/index.html.

Institute for Electrical and Electronics Engineers (IEEE). *Wireless Zone Overview*. Accessed: September 7, 2005. Available: http://standards.ieee.org/wireless/overview.html.

Jainchill, Johanna. "For Surfers, a Roving Hot Spot That Shares." *New York Times*. July 14, 2005.

Keenan, Robert. "Caching technique eases WLAN roaming." *CommsDesign.com*. (August 9, 2004). Available: www.commsdesign.com/ news/tech_beat/showArticle.jhtml?articleID=26806681.

Kenney, Brian. "The Library Reloaded." *Library Journal*. (December 15, 2003). Available: www.libraryjournal.com/article/CA339667.html.

Kewney, Guy. "Linksys falls off Wi-Fi bridge." *The Register*. (March 3, 2004). Available: www2.theregister.co.uk/2004/03/03/linksys_falls_off_wifi_bridge.

Koman, Richard. "Will Congress Ban Municipal Wifi?" *O'Reilly Policy Center*. (August 3, 2005) Available: www.oreillynet.com/pub/a/policy/2005/08/ 03/muniwifi.html.

LaRocca, James and Ruth LaRocca. *802.11 Demystified*. New York : McGraw-Hill, 2002.

Layton, Julia and Curt Franklin. "How Bluetooth Works." Howstuffworks.com. Accessed: November 12, 2005. Available: *http:// electronics.howstuffworks.com/bluetooth.htm*.

Levine, Rick. "Regional 911." *Centralities* (August 2005). Available: www.cmrls.org/centralities/2005/August.html.

"Linskys.com – Products/Wireless/Basic Networking/Print Servers/Wireless-G/ WPS54G." Linskys.com. Accessed: January 2, 2006. Available: *www.linksys.com*

Mathias, Molly Susan and Steven Heser. "Mobilize Your Instruction Program With Wireless Technology." *Computers in Libraries* 22, no. 3 (March 2002): 24.

McKeag, Louise. "WLAN Roaming—The Basics." *Techworld.com*. (March 23, 2004). Available: http://techworld.com/mobility/features/index.cfm? featureid=435&Page=1&pagePos=16.

Mobile Pipeline Staff. "Wi-Fi Security Service For Homes, SMBs Launched." *Techbuilder.org*. (March 31, 2005). Available: http://www.techbuilder.org/ news/160400491.

Muller, Nathan J. *Wireless A to Z*. New York: McGraw-Hill, 2003.

"NAS Wireless—Survey." NASWireless.com.2005. Available: *http:// www.naswireless.com/survey.html*. Accessed: October 23, 2005.

National Air and Space Administration. *Electromagnetic Spectrum—Introduction*. (1997–2005). Available: http://imagine.gsfc.nasa.gov/docs/science/ know_l1/emspectrum.html.

Newell, Bruce. "Make Your Library a Wi-Fi Hotspot." *WebJunction*. (November 30, 2004). Available: www.webjunction.org/do/DisplayContent?id=8211.

Nolan, Chris and Joshua Trevino. "Two Sides to the Municipal Wi-Fi Story." *eWeek.com*. (November 22, 2005). Available: www.eweek.com/article2/ 0,1895,1892497,00.asp.

Palmer, Michael J. and Robert Sinclair. *Guide to Designing and Implementing Local and Wide Area Networks*. Boston, Mass.: Thomson/Course Technology, 2003.

Phifer, Lisa. "Helpful Steps in Troubleshooting Your Wireless Connectivity Problem." *Techtarget.com*. (January 20, 2004). Available: http:// searchmobilecomputing.techtarget.com/originalContent/ 0,289142,sid40_gci1027249,00.html.

Piper, Dave. "Wireless Classroom Project: Arizona Health Sciences Library, University of Arizona, Tucson" (July 2001) Available: www.ahsc.arizona .edu/wireless/.

Pogue, David. "You're Ice Cold at a Hot Spot: 7 Reasons Why." *New York Times*. May 4, 2005.

Posey, Brien M. "How to Troubleshoot Your Wireless Network." *Techrepublic.com*. (November 18, 2002). Available: http://tech republic.com.com/5100–1035_11–1055909.html?tag=hdi.

Reardon, Marguerite. "New Wi-Fi standard takes the slow road." *CNET.com*. (May 19, 2005) Available: http://news.com.com/New+Wi-Fi+standard+takes+the+slow+road/2100–7351_3–5714368.html.

Reid, Neil and Ron Seide. *802.11 (Wi-Fi) Networking Handbook*. Berkeley, CA.: McGraw-Hill/Osborne, 2003.

Ross, John. The *Book of Wi-Fi: Install, Configure, and Use 802.11b Wireless Networking*. San Francisco: No Starch Press, 2003.

Rysavy, Peter. *Planning and Implementing Wireless LANs*. (January 14, 1998). Available: www.networkcomputing.com/netdesign/wlan1.html.

Sauers, Michael. "A Library Policy for Public Wireless Internet Access." *WebJunction*. (July 1, 2005) Available: www.webjunction.org/do/ DisplayContent?id=11033.

Schuyler, Michael. "The Next Big Thing: Super-PDAs Do It All." *Computers in Libraries* 22, no. 6 (2002): 28.

Schuyler, Michael. "Planning for Those 'Unknown Unknown' Projects." *Computers in Libraries* 22, no. 9 (2002): 38.

Shim, Richard. "FAQ: Wi-Fi Alphabet Soup." *CNET.com*. (April 14, 2005). Available: http://news.com.com/FAQ+Wi-Fi+alphabet+soup/2100–1041_3– 5669837.html.

Shim, Richard. "Group Proposes Faster Wi-Fi Specification." *CNET.com*. (August 12, 2004). Available: http://news.com.com/Group+proposes+faster+Wi-Fi+specification/2100–7351_3–5307332.html.

Simone, Dan. "Calculating Bandwidth Per application For Access Point Coverage." *Networkworld.com*. (March 22, 2004). Available: www. networkworld.com/columnists/2004/0322wizards.html.

Stone, Laura. "Wireless in Arizona Libraries—Bootstrapping with LSTA Funds." *WebJunction*. (July 1, 2005). Available: www.webjunction.org/do/ DisplayContent?id=11015.

"A Support Guide for Wireless Diagnostics and Troubleshooting." *Microsoft.com*. (May 31, 2005). Available: www.microsoft.com/technet/ prodtechnol/winxppro/maintain/wlansupp.mspx.

Tennant, Roy. "Digital Libraries: Free as a Bird: Wireless Networking Libraries." *Library Journal*. (August 15, 2002) Available: www.libraryjournal.com/article/CA237623.html.

"TKIP (Temporal Key Integrity Protocol)." Wikipedia.org. Accessed: December 29, 2005. Available: http://en.wikiipedia.org/wiki/TKIP

Tekinay, Sirin (Ed.). *Next Generation Wireless Networks*. New York: Kluwer Academic Publishers, 2002.

Unger, Jack. *Evaluating and Selecting Wireless Equipment*. Indianapolis: Cisco Press. (February 2003). Available: www.windowsitlibrary.com/Content/1130/06/toc.html.

"University Library Navigation Enabled by Ekahau." *Directions Magazine*. (June 12, 2003) Accessed: March 15, 2005. Available: www.directions mag.com/press.releases/index.php?duty=Show&id=7276.

U.S. Department of Commerce, National Telecommunications and Information Administration (NTIA). *United States Frequency Allocations*. (October 2003) Available: www.ntia.doc.gov/osmhome/allochrt.pdf.

U.S. Department of Commerce, National Telecommunications and Information Administration (NTIA). *Plan to Select Spectrum for Third Generation (3G) Wireless Systems in the United States*. (October 20, 2000). Available: www.ntia.doc.gov/ntiahome/threeg/3g_plan14.htm.

Walker, Jonathan. "Wi-Fi Mesh Networks, The path to mobile Ad Hoc." *Wi-Fi Technology Forum Tutorial*. (2004). Available: www.wi-fitechnology.com/Wi-Fi_Reports_and_Papers/Introduction_to_Wi-Fi_Mesh_Networks.html.

Wearden, Graeme. "Wi-Fi Group: Jumping the Gun on Gear is Risky." *CNET.com News*. (October 12, 2004). Available: http://news.com.com/Wi-Fi+group+Jumping+the+gun+on+gear+is+risky/2100−7351_3−5406664.html.

Wearden, Graeme. "Wi-Fi Testing Finds Weak Links." *CNET News.com*. (January 12, 2004) Available: http://news.com.com/2100−7351_3−5139499.html.

Wikoff, Karen. "Should Your Library Go Wireless? Resources for Deciding" (August 2001). Available: www.geocities.com/karinwikoff/IST511Wireless.htm.

Williams, Joe. "Taming the Wireless Frontier: PDAs, Tablets, and Laptops at Home on the Range." *Computers in Libraries* 23, no. 3 (2003): 10.

Williams, Robert L. "Wireless Community Networks: A Guide for Library Boards, Educators, and Community Leaders." Texas State Library and Archives Commission. (April 26, 1999). Available: www.tsl.state.tx.us/ld/pubs/wireless/.

"Wireless Andrew History". Pittsburgh: Carnegie Mellon University. Available: www.cmu.edu/computing/wireless/wirelesshistory.html.

"Wireless Group Says Interoperability Problem Rising." *Associated Press/CRN*. (March 18, 2004). Available: www.crn.com/article/link.jhtml?articleId=18841118.

"WLAN—Roaming." *UNINETT*. (April 15, 2005). Available: www.uninett.no/wlan/roaming.html.

"WPAN." *Techtarget.com*. (March 14, 2004). Available: http://searchmobilecomputing.techtarget.com/gDefinition/0,294236,sid40_gci837444,00.html.

JOURNALS TO WATCH

Computers in Libraries magazine from Information Today (www.infotoday.com) is a great resource for a number of library technology issues. In particular, their March 2003 issue, "Welcoming Wireless to Your World," is wonderfully helpful as a primer. We also recommend articles by Marshall Breeding on wireless in libraries.

InfoWorld. Has a "Special Reports" section on managing wireless networks. (www.infoworld.com)

NetworkWorld. A popular IT trade journal, that includes sections on wireless LANs, security, etc. It is available on the Web; look for the "Wireless/Mobile" section. (www.networkworld.com)

WEB SITES AND E-MAIL LISTS

Library Technology Guides are edited by Marshall Breeding, the Library Technology Officer for the Jean and Alexander Heard Library at Vanderbilt University, whose articles for *Computers in Libraries*' "Systems Librarian" column are aggregated here. RSS feed is available for updates. (www.librarytechnology.org).

Libwireless. This highly recommended moderated e-mail list covers wireless issues in libraries. To subscribe, use either of these URLs: http://people.morrisville.edu/~drewwe/wireless/libwireless.html or http://ls.suny.edu/read/all_forums/subscribe?name=libwireless.

Syslib subscription information for this email list on Systems Librarianship is available at http://listserv.buffalo.edu/archives/syslib-l.html.

Web4Lib e-mail list. This highly recommended and long-standing e-mail list on Web-related topics in libraries is available via WebJunction. Many of the user experiences contained in this book came from Web4Lib members. To subscribe go to: http://lists.webjunction.org/web4lib/.

WebJunction. This highly recommended library-focused technology help site is an excellent starting point. Although it is partly focused on Gates computers (it is funded by a Gates grant), it contains good general information useful to all librarians. If your state library doesn't already subscribe, encourage them to do so at: www.webjunction.org. In particular, the Wireless Discussion Board is very fruitful and a great place to search for answers and post questions (see http://webjunction.org/forums/forum.jspa?forumID=44). For resources related to wireless see http://webjunction.org/do/DisplayContent?id=11561.

Webopedia. When you're reading documentation on wireless hardware, you may encounter new jargon and acronyms. This online dictionary of computer and Internet-related terms is a great, quick spot to find definitions. (www.webopedia.com)

Wi-Fi Alliance. This industry consortium of companies involved in wireless product development, promotes wireless technology and its uses, and also certifies interoperability of new wireless products. The phrase "Wi-Fi certified" comes from the Wi-Fi Alliance. Its Web site has excellent, clear explanations of what Wi-Fi is, how it works, information on Wi-Fi security, and more. It is highly recommended, especially for product information. (www.wi-fi.org)

Wi-Fi Planet. Owned by Jupitermedia Corp., this Web site focuses on wireless networking. It provides good articles, including white papers on new issues. (www.wi-fiplanet.com)

Wireless Librarian Blog. Discussion from Bill Drew's *Wireless Librarian* Web site was moved to this multi-user blog. This excellent resource for questions on wireless is available as RSS feed. (www.wirelesslibraries. blogspot.com)

Wireless Librarian Owned by Wilfred (Bill) Drew, this Web site includes a list of Wi-Fi site finders, so you can find libraries with wireless.http:// people.morrisville.edu/~drewwe /wireless/index.htm.

WLANA. Wireless LAN Association describes itself as "a non-profit educational trade association, comprised of the thought leaders and technology innovators in the local area wireless technology industry." (www.wlana.org)

TROUBLESHOOTING HELP WEB SITES

Apple Users—Airport Support Pages. www.apple.com/support/airport/.

"How to Set Up Your Computer for Wireless Networking in Windows XP." *Microsoft.com.* (November 5, 2004). Available: http://support.microsoft.com/ default.aspx?scid=kb;en-us;314897.

"How to Troubleshoot Wireless Network Connections in Windows XP." *Microsoft.com.* (June 22, 2004). Available: http://support.microsoft.com/ kb/313242/.

"MSBBN: How to troubleshoot wireless connection problems." Microsoft.com. (February 8, 2005). Available: http://support.microsoft.com/default.aspx? scid=kb;en-us;831770.

Palm Users—Palm/Treo Support Pages. www.palm.com/us/support/.

"A Support Guide for Wireless Diagnostics and Troubleshooting." *Microsoft.com.* (May 31, 2005). Available: www.microsoft.com/technet/ prodtechnol/winxppro/maintain/wlansupp.mspx.

"Troubleshooting Windows XP IEEE 802.11 Wireless Access." *Microsoft TechNet.* (January 26, 2005). Available: www.microsoft.com/technet/ prodtechnol/winxppro/maintain/wifitrbl.mspx.

"Understanding 802.1X Authentication for Wireless Networks." *Microsoft TechNet.* (January 21, 2005). Available: www.microsoft.com/technet/ prodtechnol/windowsserver2003/library/ServerHelp/908d13e8-c4aa–4d62– 8401–86d7da0eab48.mspx.

"Westchester Libraries Wireless FAQ." Available: www.westchesterlibraries.org/ wireless_print.html.

WIRELESS TECHNOLOGIES

Bluetooth. www.bluetooth.com

IEEE 802 LAN/MAN Standards Committee. Discusses all approved and in development standards under the 802 heading. Good to keep an eye on for new developments, especially in the areas of wireless security (802.1x, etc.). (www.ieee802.org).

Worldwide Interoperability for Microwave Access (WiMAX)
www.wimaxform.com
www.intel.com/netcomms/technologies/wimax
www.wi-fiplanet.com/wimax

SITE SURVEY AND TESTING TOOLS

Canary Wireless. Their Digital Hotspotter ® product is a small device that can detect Wi-Fi signals and display minimal information on a small LCD screen; including SSID, signal strength, channel and encryption status. Similar products are sold by several vendors, including Targus, PCTEL, Mobile Edge, and others. Canary Wireless' product most consistently receives good reviews. (www.canarywirless.com)

NetStumbler. A free software tool for Windows, used widely, which allows you to detect WLANs. Load it on a laptop and move around the testing area to find "rogue" signals or to test the efficacy of access point placement. (www.nestumbler.com)

CAPTIVE PORTALS—SOFTWARE AND HARDWARE SOLUTIONS

The following list is not comprehensive, simply a sampling of available solutions. We highly recommend talking to your wireless vendor about options they use, and in particular the solutions that are associated with the hardware you're considering purchasing.

Aptilo Access Gateway: www.aptilo.com/solutions_public_access.htm

Cisco BBSM Hotspot: www.cisco.com

Cisco Site Selection Gateway (SSG) / Subscriber Edge Services (SESM): www.cisco.com

Nomadix Service Engine: www.nomadix.com/products/

Personal Telco Wiki: http://wiki.personaltelco.net/index.cgi/PortalSoftware

Wikipedia.org: http://en.wikipedia.org/wiki/Captive_portal

WI-FI SECURITY

Wi-Fi Alliance. See the following section of the Wi-Fi Alliance Web site for clear explanations of basic wireless security options and terms. (www.wi-fi.org/OpenSection/secure.asp?TID=2)

LIBRARY-SPECIFIC WIRELESS PRODUCTS AND SERVICES

AirPac by Innovative Interfaces: www.iii.com/mill/webopac.shtml
Bluesocket BlueSecure Controller: www.bluesocket.com/
Polaris Wireless Access Manager: www.polarislibrary.com/products_Services/ wam_info.asp
PublicIP's ZoneCD for Library Hotspots: www.publicip.net
Sirsi PocketCIRC: www.sirsi.com/Pdfs/Products/SirsiPocketCirc.pdf
TLC WirelessSolution: www.tlcdelivers.com/tlc/pdf/wirelesssolution.pdf

SOURCE C: SAMPLE POLICIES AND FAQS

WIRELESS ACCESS POLICIES

Before you implement any new technology or system, it's a good idea to review your existing policies and verify if you need to augment to them. Below you will find a review of the Wireless Access Policy checklist from Chapter 3, which gives you a list of items to think about in reference to creating a wireless policy. You will also find references to articles about creating a wireless policy in a library. URLs of libraries with published wireless access policies and also some examples of those policies, with our comments. The examples are at the end of the section, by figure number. Don't forget to look at the original policies on the Web, as they may have changed or expanded since we took screenshots.

ARTICLES
DeGroff, Amy Begg. "Howard County Library WiFi Policy." *WebJunction*. (July 25, 2005).
Available: www.webjunction.org/do/DisplayContent?id=11059.
Sauers, Michael. "A Library Policy for Public Wireless Internet Access." *WebJunction*. (July 1, 2005).
Available: www.webjunction.org/do/DisplayContent?id=11033.

WEB SITES WITH LISTS OF POLICIES
Good Internet policy lists at *WebJunction*:
www.webjunction.org/do/Navigation?category=394
Best Practices in Public Libraries (includes links to some good Internet and Wireless Use Policies): www.ala.org/ala/pla/resources/bestpractices.htm

INDIVIDUAL LIBRARIES' POLICIES
Austin (TX) Public Library's Acceptable Use Policy for Wireless Access: www.ci.austin.tx.us/library/wireless.htm
Besides a catchy wireless logo, Austin's policy is clearly laid out and covers a lot of ground. (see Figure C–1) You have to go over to its FAQ ("Wireless @ APL" linked at bottom) to find out that you have to register (through a gateway) before you can start surfing.

Chicago (IL) Public Library's Wireless Policy: www.chipublib. org/003cpl/computer/wifi/wifi.html.

It was interesting to see that many of the larger libraries had fairly small policies. Simple, broad disclaimers were more the norm, followed by very clear instructions in their FAQs as to how and where to access wireless, often with links to an acceptable use policy covering all activity within the library. (See Figure C–2)

Monterey (CA) Public Library's Internet Policy: www.monterey. org/library/ipolk.html

They have an interesting "explanation of policy for kids," which we haven't seen elsewhere. (see Figures C–3 and C–4)

Hamilton College's Wireless Policy: www.hamilton.edu/college/ its/policies_standards_plans/wireless_policy.html

Good example of a policy from an academic library.

C.E. Weldon Public Library (Martin, TN) Wireless Policy: www.ceweldonlibrary.org-wireless.htm

Includes a shot of the wireless network connection page in XP, with the library's SSID showing. (See Figure C–5)

Iowa State University's Wireless Policy: http://policy.iastate.edu/ it/wireless/

This extensive and excellent policy covers a lot of ground.

Morton Grove (IL) Public Library—Webrary®—Wi-Fi Access Policy: www.webrary.org/inside/polwifi.html

This library's policy requires users to "click-through" a user agreement before using wireless.

Pikes Peak (CO) Library District—About Your Library—Wireless Access: http://library.ppld.org/AboutYourLibrary/Wireless/ default.asp

This library requires a library card number and PIN and seem to be using a captive portal. The actual policy is a link from the page above, in PDF format.

City of Walla Walla, WA Policy—sample of policy points: www.mrsc.org/govdocs/w33-epol.aspx.

One of the most comprehensive policies we've ever seen.

West Deptford (NJ) Free Public Library – Wireless Network Access Policy: www.westdeptford.lib.nj.us/wireless_policy.htm

This basic policy warns, "The library's network is not sized for large downloads from the Web." (See Figure C–6)

SOME ARTICLES ON LIBRARIES THAT ARE CIRCULATING LAPTOPS

Allmang, Nancy. "Our Plan for a Wireless Loan Service." *Computers in Libraries* 2,3 no. 3 (2003): 20.

Allmang, Nancy. "Wireless Loan Service—Updating Our Story." Letter to the Editor. *Computers in Libraries* 25, no. 7 (2005): 50–52.

POLICIES OF LIBRARIES THAT ARE CIRCULATING LAPTOPS

Bucks County Community College Library Laptop Computer Policy: www.bucks.edu/library/services/laptop.html.

Of the laptop checkout policies we found, BCCC's seemed to be a generous policy with clear guidelines. Like many libraries, it requires students to sign a once per year user agreement (see Figure C–8) before checking out a laptop. At the moment the library does not allow students to connect with their own laptops, though it are moving in that direction. (See Figures C–7 and C–8)

We have been lending laptops for three semesters now and it has gone quite smoothly. The service has been well received by the students; they do not mind completing the borrower's agreement or handing over their driver's license in addition to their student ID when borrowing a machine.

There is some discussion about modifying the policy to allow laptops to circulate out of the library for short periods of time, like two or three days. This is just a discussion at this point.

What our students would really like is an open wireless network so that they may connect to the Internet with their own machines. I know our technology staff is working to that end, as well as providing a way for students to access printers from their own machines. It is my understanding that there exist many network tasks in order to achieve this while maintaining a secure system. So for now, we wait.

Brian T. Johnstone
Public Services Librarian
Bucks County Community College
Newtown, PA

Tarleton State University's Laptop Computers Policy: www.tarleton.edu/~library/laptops.html.

The Dick Smith Library's policy and FAQ page includes maps of where connected printers are located on each floor. As the laptops are maintained by the library and IT department, they can be configured for network access, and thus for printing. This is common for circulating laptops, and gives some value-added aspects (printing, university network access, etc.) to checking out a laptop as opposed to using one's own. This is balanced by the obvious limitations of time and possible fines.

Old Dominion University Libraries' Laptop Policy: www.lib. odu.edu/services/computing/laptoppoliciesprocedures.htm.

Some items to note on this fairly standard example of a laptop checkout policy: unlike the University of New Brunswick Libraries Old Dominion offers a power cord for patron use, and does not require battery use only. It also offers network access for saving documents, but via a Web-based network access solution, which is a good security choice.

University of New Brunswick Libraries Laptop Checkout Policy: www.lib.unb.ca/about/policies/wireless_policy.html.

See also the Wireless FAQ relating to circulating laptop use only: www.lib.unb.ca/wireless/wirelessFAQ.html.

UNB Libraries have an extensive policy and separate FAQ. Their limitation to "battery use only" may be due to limited outlets available throughout the library. This is not uncommon. Circulating laptops and providing wireless access for them provides some flexibility. They also allow outside laptops, as long as they are pre-registered by students and faculty. (See Figure C–9)

Austin (TX) Public Library Sample FAQ: www.ci.austin.tx.us/ library/wireless_at_apl.htm.

Again, the catchy logo is great. The registration explanation could use a screenshot or two for clarity. The highlighting and boxing off of troubleshooting steps is eye-catching. (See Figure C–10)

Boulder (CO) Public Library Sample FAQ: http://www.boulder. lib.co.us/general/wireless_faqs.html.

Good Q&A format here. Each question has a jump-down link, which is useful since the page is quite long. The troubleshooting info should be on a separate page, but with the jump links, it works allright. (See Figure C–11)

Chicago (IL) Public Library Wireless FAQ: www.chipublib.org/ 003cpl/computer/wifi/wififaqs.html.

This is an extensive and useful FAQ. That uses the Q&A format and includes an explanation of 802.11 types and also an acronym listing. That's really thinking ahead to what the patron might need. There is also a good explanation of why SMTP e-mail will not be available. Many libraries block this, but don't tell patrons until they ask, giving them poor explanations of why. It's a perfectly sound decision for security, so don't feel bad about giving an explanation. (See Figure C–12)

Penn State George T. Harrell Library Wireless FAQ: www.hmc.psu.edu/library/GenInfo/wirelesspolicy.htm

Includes information on checking out wireless network cards and the pay for print system that students can use, which requires a script to be run on your laptop. (See Figure C–13)

St. Joseph County (IN) Public Library Sample FAQ: http://sjcpl.lib.in.us/services/wifi/docs/howtowifi.pdf.

A simple, straightforward troubleshooting sheet, in PDF format.

Westchester (NY) Libraries Sample FAQ: www.westchesterlibraries.org/wireless_print.html.

Very simple, but gets the point across with nice links to all member libraries who have wireless.

Peabody Library Sample FAQ: www.peabodylibrary.org/reference/wireless_faq.html.

This page has a shot of the "wireless login screen" users will see when they login which is a good idea.

Marin County (CA) Free Library Sample FAQ: www.co.marin.ca.us/depts/lb/main/policies/InternetFAQ.cfm.

This is a huge FAQ with (not wireless specific, per se) each question jump-linked to its answer. Although a bit cumbersome, this site covers a lot of ground actually tackling some filtering policy issues and a lot of general information about using the Internet.

New Milford Public Library Sample FAQ: www.biblio.org/newmilford/Working/wireless.htm.

With step-by-step connection and troubleshooting instructions, as well as screenshots of screens that patrons will see, this site uses a portal requiring library card authentication.

Roselle Public Library, Roselle, Illinois (IL) Sample FAQ: www.roselle.lib.il.us/GeneralInfo/ComputerResources_WirelessFAQ.htm.

This FAQ is an example of an excellent Q&A format.

Figure C–1   Sample Wireless Policy. Austin (TX) Public Library.

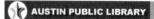

Austin Public Library
Acceptable Use Policy
for Wireless Access

I understand and agree to comply with all of the following conditions:

- I will comply with all state and federal laws and the Austin Public Library policies and procedures.
- I understand and acknowledge that the Internet contains information, both written and pictorial, which may be offensive or harmful to me or to others.
- I release the City of Austin and the Library from all liabilities associated with my viewing of, use of, or exposure to any information, machine-readable file, picture, graphical representation, or illustration I may encounter while using any public-use workstation, whether or not such information appears or is delivered through the station I operate.
- I will not harm or harass any City employee or member of the public.
- I will not violate any state or federal statute including those regarding obscenity, pornography, or delivery to minors material deemed harmful to them.
- I understand that I create, store, and use personal files at my own risk and that the Library is not responsible for the loss of personal electronic documents, diskettes, and/or files.
- I understand and accept that my failure to comply with Library policies and procedures will result in Library penalties that range from suspension of Internet privileges for a week through permanent eviction from the Library.
- I understand that the City of Austin reserves the right to change its public use workstation access policies and procedures.
- I understand that individual Library staff members are not authorized to modify these policies.

Return to Wireless @ APL.

Austin City Connection - The Official Web site of the City of Austin
Contact Us: Send Email or 512-974-7400.
Legal Notices | Privacy Statement
© 2001 City of Austin, Texas. All Rights Reserved.
P.O. Box 1088, Austin, TX 78701 (512) 974-2000

Figure C–2   Sample Wireless Policy. Chicago (IL) Public Library.

Chicago Public Library

Home | Search our Catalog | Find It! | View your library card record | Index

WiFi

FREE WiFi COMPUTING AT THE CHICAGO PUBLIC LIBRARY

Welcome to the Chicago Public Library's High Speed Wireless Internet System. Free access is provided in nearly all 79 Chicago Public Library locations. Getting online is quick and simple. All you need is a wireless enabled laptop computer, tablet PC or PDA. The Library's network is open to all visitors free of charge and without filters. No special encryption settings, user names or passwords are required.

Hardware Requirements

A laptop computer with a WiFi Internet card that supports the WiFi standard (also known as IEEE 802.11b or g).

Quick Setup Settings:

SSID (network name) = CPLWiFi
Test your connection. If connection fails, you may need to adjust the following settings:
WEP = disable WEP encryption
Mode or Network Type = Infrastructure mode or Access Point
DHCP = Obtain an IP address automatically
DNS = Obtain DNS server address automatically

Troubleshooting
This section includes Windows 2000, XP, NT and Macintosh network settings. (Image edited for size)

Limitations and Disclosures

The actual policy section is small, but clear.

They do require the acceptance of a user agreement before using.

The Library's wireless network is not secure, and the Library cannot guarantee the safety of your traffic across its wireless network. The Library assumes no responsibility for the configurations, security or files on your laptop resulting from connection to the Library's network. Information sent to or from your laptop can be captured by anyone else with a wireless device and appropriate software, within up to three hundred feet.

The Library is not able to provide technical assistance to you, and there is no guarantee that you will be able to make a wireless connection. If you need assistance, contact the manufacturer of your laptop or software. Some helpful hints are provided on the Chicago Public Library's Web site. The Library is not responsible for any changes you make to your computer's settings.

You must click the I AGREE button to connect to the Library's wireless network. If you successfully connect, you will be taken to the Chicago Public Library's website.

For more information see our WiFi FAQ's page.

◀ Back to the Chicago Public Library Home Page

Read, Learn, Discover! 24 HOURS A DAY

Figure C–3   Sample Wireless Policy for Adults. Monterey (CA) PL.

CITY OF MONTEREY

Excerpt from: www.monterey.org/library/ipola.html

Monterey Public Library's Internet Policy

Complete Internet Access and Use policy
Explanation of the rules for kids ⟵
Previous Page

Again, their inclusion of an interpretation "for kids" is unusual and noteworthy. (See next image)

In order to use Monterey Public Library's Internet service, you must follow the rules established by the Library Board of Trustees. If you do not follow these rules, you will not be permitted to use the Internet and you may be required to leave the Library.

Monterey Public Library does not monitor, control, sponsor or endorse the material you find on the Internet. The Library does not edit or restrict Internet information. Only you and your family have the right and responsibility to define what material or information is consistent with your personal and family beliefs.

It is worth looking at this whole policy online (we have edited the image for size). Clearly, this policy took considerable time and effort. It is extensive and covers a number of contingencies.

Both wireless-specific and general Internet use policies/disclaimers are included.

The Library cannot guarantee your privacy when using the Internet. Respect the privacy of other Internet users, and do not attempt to show displayed material to passersby. Library staff must take appropriate actions to resolve problems which arise during use of the Library's Internet service and to enforce Library policies and rules. To this end, Library staff members may need to observe Internet use, question Internet users, and restrict conduct by Internet users which violates this policy.

The wireless network is not encrypted. Other wireless users may intercept any information you send or receive using the wireless network, and web-based security controls may not be sufficient protection. Therefore, the Library recommends that you do not enter credit card numbers, passwords, or other personally identifiable information.

Other wireless users may also be able to view or change files on your computer. The Library recommends that you install and use anti-virus software, firewall software, and current security patches or upgrades on your computer.

The Library makes no representation or guarantee that any part of the Internet service, including the wireless service, will be uninterrupted, error-free, virus-free, timely, or secure, nor that any Internet content is accurate, reliable or safe in any manner for download or any other purpose.

You must follow Library rules while using the Internet (ask to see Policy No. 515, Disruptive Behavior in the Library). You must be quiet, courteous toward others and respectful of Library equipment.

Rev. 10/27/2005 D. Holtzman http://www.monterey.org/library/ipola.html

Figure C–4   Sample Wireless Policy for Kids. Monterey (CA) PL.

Internet Rules ⟶

Monterey's Internet Rules "For Kids"

To use the Monterey Public Library's Internet service, you must follow the rules. These rules are written for kids. You can also read the rules written for <u>ADULTS</u> or the Library's whole <u>INTERNET POLICY.</u> If you don't follow the rules, you will not be able to use the Internet, and you may be told to leave the Library.

The Library's Internet service is mostly for learning. The staff gives help and suggests links, but the Library is not in charge of the Internet. In fact nobody is in charge of the Internet.

You and your family are in charge of your Internet searches. Please use the Internet in the way your family decides is right for you. In addition to the many wonderful things on the Web, there are scary and disgusting things too. However, if you stay on the youth pages or ask for help, you should be safe. But, again, you and your family must decide what is right for you.

> We like the way they explain this.

Library staff members want to help you to stay safe and to find what you want, but they do not have the time to teach you everything about the Internet. Please ask for help, but understand if we can't stay with you while you work.

While you are using the Internet, you are expected to do these things:

1) Limit yourself to 30 minutes, if someone is waiting.

2) Follow all other Library rules, be quiet and courteous to others, and take good care of Library furniture and equipment.

3) No chat is allowed. This includes instant messaging.

4) No "hacking." You cannot change the way the Library's Internet service is set up. You cannot download or bookmark or use your own software. You cannot go into Library files. (You can suggest links by using the <u>Suggestion Box</u>.)

5) Respect the privacy of others. Do not show your screen to others and do not look at other people's work. Keep in mind that the Library is really not a good place for private work.

6) Respect copyright laws. It may be against the law to copy some things on the Internet.

7) Don't break the law. Some places on the web are for adults only. It would be against the law to go there. It would also be against the law to lie about your age or who you are.

Figure C–5   Sample Wireless Policy. C.E. Weldon Public Library (Martin, TN).

C.E. WELDON
PUBLIC LIBRARY
A Tradition of Information Service

Home
Director's Message
About the Library
Children's Library
Library Services
FAQ
Resources & Links
Replicas
Upcoming Events
Teen Page
Donations

Excerpt from: www.ceweldonlibrary.org/wireless.htm

Wireless Internet Access

Free wireless Internet access is available at C. E. Weldon Public Library. You don't need a plug or phone jack, just turn on your notebook/laptop computer or other wireless device and start surfing.

Here they include some minimal technical requirements
and their limitations and disclaimers. (Image edited for size.)

How to get connected:

Select CEWELDON as the wireless access point. Other wireless networks may be available.

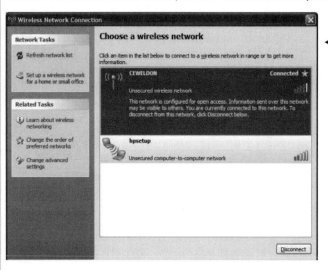

We readily applaud
the use of images
and screenshots to
explain connection
issues.

Especially shots that
show the actual
SSID or other library-
specific information
the patron will see.

Please note:

By using wireless technology, it is possible for personal information to be compromised while being transmitted. The library will not accept responsibility for any loss or damages and, as such, encourages its patrons to utilize secure services when accessing resources in a wireless environment.

C. E. Weldon Public Library

100 Main Street · Martin, Tennessee 38237 · (731) 587-3148 FAX (731) 587-4674 · weldonlib@charter.net

Library Hours: Monday - Friday: 9:30 AM — 5:30 PM, Thursday: 9:30 AM - 8:00 PM, Saturday: 9:30 AM — 12:00 PM, Sunday: Closed

Copyright © 2003 C. E. Weldon Public Library. All Rights Reserved.

Figure C–6   Sample Wireless Policy. West Deptford (NJ) Free Public Library.

West Deptford Free Public Library
420 Crown Point Road, Thorofare, NJ 08086 (856) 845-5593

Technology Services
Wireless Network Access Policy

Excerpt from:
www.westdeptford.lib.nj.us/wireless_policy.htm

The wireless network in the West Deptford Free Public Library was implemented in October 2002 for patrons who wish to use their own laptop when at the library. All terms and conditions set forth in the West Deptford Free Public Library's Library's Internet Use Policy and Technology Policy are applicable to wireless network access.

A patron's laptop must be configured with virus protection using current virus definitions. Patrons are expected to know how to configure their own laptop for wireless use. The library does not offer wireless access support. The Circulation Desk can provide a document with technical details and an example of configuring Windows XP for wireless access.

Most WiFi (802.11b and 802.11g) network cards will be compatible. However, the West Deptford Free Public Library can make no guarantees as to compatibility of patron's equipment with the library's network.

The West Deptford Free Public Library wireless network has been configured and sized for interactive searches and knowledge gathering on the World Wide Web. The library's network is not sized for large downloads from the web. Therefore, the downloading of software, large images, music, etc from the web onto a patron's laptop is not permitted. Patrons will be asked to immediately terminate any activity that adversely impacts the library's network performance.

June 2003

The FAQ below has been excerpted.
See the URL above for the entire document.

Wireless Internet Access
in South Jersey Public Libraries

| Home / FAQ | About The Project | Participating Libraries | Technical Assistance |

We are pleased to offer free wireless Internet access to customers with appropriately equipped wireless devices.

Answers to Frequently Asked Questions (FAQs)

 What can I do on the wireless network?   ← *The Q&A format is very engaging.*

 Our wireless network offers you have a high speed connection to the Internet that let's you:

- Check your e-mail
- Surf the Internet
- Do online research
- Check the library catalog
- Access the library's databases

← *Don't just tell them what they cannot do, tell them what they can do!*

Printing is not currently available from the wireless network. To print you can either save your document and print when you get home, or save your document to a disk or cd-rom and use one of the other computers available to you in the library.

 Can the Library help me configure my computer?   *Link to technical support pages.*

 Each computer is unique, and we're sorry that we are not able to offer personal technical assistance with your configuration. However, we do provide a webpage of technical assistance to help you with the most common issues. The technical assistance webpage is available at: www.sjrlc.org/wireless/tech.htm

 Which South Jersey public libraries offer wireless hotspots? ←   *Offer them options when available.*

 Click Here for a complete list of libraries that are offering wireless hotspots.

Figure C–7 Sample Wireless Laptop Lending Policy. Bucks County Community College (PA).

BUCKS COUNTY COMMUNITY COLLEGE

About Bucks | Academics | Admissions | Arts & Culture | Center for Workforce Development

Continuing Education | Online Learning | E-Services | Library | Student Life | Student Services

Search

[   ] [>]

[Quick Links]

Library
Laptop

www.bucks.edu/library/services/laptop.html

Wireless Laptop Lending Policy

This section includes the eligibility requirements--faculty, staff and students with patron account in good standing. (Edited for space.)

BORROWING GUIDELINES

- Wireless laptops are for use only in the main library on the Newtown Campus, floors 2 through 4.
- Check out period is for three hours; students shall be allowed one renewal for three hours.
- Laptops are available on a first-come, first-served basis.
- Usage time shall coincide with Library hours, terminating fifteen minutes before library closing time.
- Borrowers are solely responsible for laptop during the check out period, including damage, loss, and theft.
- Borrowers shall not leave laptop unattended while checked out. **——> Good reminder!**
- All file saving shall be the responsibility of the user, shall be to a disk or flash memory, and no files shall be saved to the laptop.

Compared to some other laptop checkout policies we saw, BCCC's was quite generous as to time.

BORROWING PROCEDURE

- At the time of check out a borrower must present a valid college ID card AND a valid photo ID, both of which shall stay at the library Circulation Desk while laptop is lent out. Also, the transaction shall be recorded in the library system just like any other lending transaction; laptops shall be charged to the user account just like a book, and similarly discharged upon return.

Pre-configured network laptops allow for secure printing options, as access can be controlled.

FEATURES

- All wireless laptops shall have access to the internet and Library printers.
  - Printing is available at $0.10 per page for black & white or $0.75 per page for color.
- All wireless laptops will have the Microsoft Office suite installed.

FINES AND FEES

- Borrowers with overdue laptops shall be charged **$10.00** per hour.
- Loss of laptops shall result in a charge of **$1600.00**.
- Damage to laptop and peripherals shall result in a charge to cover repair or replacement.

Laptop fine policies we viewed did vary a bit. Generally they included a per hour charge and a significant loss charge.

Wireless Laptop Borrowers Agreement

See next Figure.

Available in the library or as a pdf here

Questions? Comments? Send us some Email

Copyright © 2004, 2005 Bucks County Community College. All rights reserved.
275 Swamp Road, Newtown Pennsylvania 18940 · 215-968-8000

Figure C–8 Sample Wireless Laptop Borrowers Agreement. Bucks County Community College (PA).

Bucks County Community College Libraries

Wireless Laptop Borrowers Agreement

This Agreement must be completed and kept on file in the Library. It will be valid for the duration of the current academic semester.

Instructions:

1. Read the Wireless Laptop Lending Policy and Guide to Responsible Use of Electronic Communication at BCCC.
2. Initial each of the rules and regulations regarding the borrowing of this equipment to indicate you understand and agree to abide by them.
3. Sign and date the bottom of this agreement.

_____ I understand that all laptops and related equipment are the property of Bucks County Community College.

_____ I certify that I am a currently enrolled student, faculty, or staff member of the College with a valid patron account in good standing.

_____ I understand that I may use the laptop only in the main library on the Newtown Campus, floors 2 through 4.

_____ I will never leave the laptop unattended.

_____ I understand that both my college ID card AND a valid photo ID will be held at the library circulation desk for the entire duration of each borrowing period.

_____ I understand that the replacement fee for lost laptops is $1600.00.

Moreover:

My signature indicates I have received, fully read, understand, and agree to abide by all the rules and regulations of the Wireless Laptop Lending Policy and the Guide to Responsible Use of Electronic Communication at BCCC; I accept responsibility for proper care of the equipment I am borrowing and understand I will be held accountable for all replacement or repair costs in the event of loss or damage; I agree to return laptop and its peripherals in good condition by the time they are due.

_____   _____

Signature   **Date**

_____

College ID Number

Figure C–9   Sample Laptop Checkout Policy and Wireless FAQ. U. of New Brunswick (Canada).

HOME > About UNB Libraries > Wireless Policy

| About | Catalogues | e-Resources | Library Units | Services | Requests | Search | Help? |

UNB Libraries Wireless Project

www.lib.unb.ca/about/policies/wireless_policy.html

Laptop Sign-out and Use Policy

(This is an excerpt. See URL above for full document.)

Borrowers:
Laptops will be available from the Circulation Desk as 2-hour use-in-library Reserve Material only. Students MUST also surrender their University Card for the duration of the sign-out period. Cards will be returned upon return of the laptop. Laptops may not for any reason leave the libraries. Laptops must be returned to the Circulation Desk within 2 hours of signout or prior to the closure of the library, whichever occurs first. Laptops may not be renewed.

Fines:
Fines for overdue laptops will be applied according to UNB Libraries' Circulation Policy. Fines will accrue at the rate of $1.00 per hour or any portion of the hour. However, unreturned laptops will be considered lost at the closure of the library. The patron will be billed and the bill will be sent the next regular working day. The bill will include the reserve fine, a $100.00 non-refundable , non-negotiable laptop fine and a replacement cost of $2500.00. In such cases, University Cards will not be returned until the matter is resolved, which would normally be the next regular working day, i.e. Monday - Friday, 8:00 am to 4:30 pm. The Registrar's Office may withhold diplomas, grades and transcripts until all outstanding library charges are paid.

Batteries: ⟶ **We did not see this limitation in very many policies.**
Patrons will not be issued power cords with the laptops. Should the battery run low during the 2 hour loan period, patrons may save their work to the network and then return to the Circulation Desk to exchange the battery.

Novell Accounts:
Patrons using the laptops are required to have UNB Fredericton or STU Novell accounts and to log onto either the UNB Fredericton or STU networks when the laptop is started.

Additional Policies, Terms and Conditions:
Patrons signing out laptops are required to abide by the Policies, Terms and Conditions as specified by the Integrated Technology Services Department of UNB Fredericton.

Wireless Networking Areas of Coverage:
Coverage maps will be available from the Circulation Desk with the laptops indicating the areas of the Harriet Irving Library that have wireless networking coverage. Science and Engineering Libraries should generally have complete coverage.

Support:
Library staff will not provide any technical support for users, apart from supplying and recharging batteries. See our Wireless FAQ for support details.

Send comments to library@unb.ca
January 2002

The FAQ linked here is extensive and includes the following items of note:

● **Wireless Coverage Maps** ⟶ **As previously noted, maps are extremely useful for wi-fi users.**

There is full wireless coverage within both the Science and Forestry, and Engineering Libraries.

Within Harriet Irving Library, however, reliable coverage is currently available **ONLY** in areas indicated on the three maps below. While there may be additional and overlapping coverage in others areas of the building, we do NOT guarantee or support access in these areas. Harriet Irving Library coverage is as follows:

HIL First Floor | HIL Third Floor | HIL Fourth Floor

From: www.lib.unb.ca/wireless/wirelessFAQ.html

>> top of page

As they control the laptops, they can control the network access configurations, and thus safely provide printing. They also provide wireless access to student and faculty laptops, but they must be pre-registered (MAC address) to connect to the network.

● **Printing**

UNB Students
Printing using the laptops is **NOT** done using the same system as the rest of the library. You **CANNOT** use the debit card system to print using the laptop. As the laptops function as student lab machines and operate on the student network, printing is done the same way it is done in the computers labs. This means that you must have Print Credits on your University account to print. (UNB : How do I get Print Credits and how much do they cost?) (STU : What do I need to know about printing?)

Choose **PRINT** as you normally would from the file menu. The proper local grayscale printer should already be set as the default, but it is a good idea to check. Collect your print job from the printer. Be sure not to accidentally grab someone else's print job.

- **Harriet Irving Library** - default printer is **HIL-CIRCDESK** (located on Circulation desk)
- **Science and Forestry Library** - default and only printer is **HIL-SCIENCE**
- **Engineering Library** - default and only printer is **HIL-ENGINEERING**

If a print job fails to appear after an excessive amount of time, see the Circulation staff. Sometimes they can fix the problem locally, but generally they will have to notify ITS.

Figure C–10   Sample FAQ. Austin (TX) Public Library.

AUSTIN PUBLIC LIBRARY   catalog | locations | reference | youth | news | links

Excerpt from
www.ci.austin.tx.us/library/wireless_at_apl.htm
(See web site for full document.)

Wireless
Austin Public Library
@ APL

**Austin Public Library
Expands Access
to Information**

Free Wireless Internet access is available at all Austin Public Library locations and
Wooldridge Square at 900 Guadalupe.

GET STARTED:

1. Bring your laptop or PDA with a IEEE 802.11b-compliant wireless networking card
   to the library.
2. Open your Internet browser. The wireless card should detect the signal
   immediately (some wireless card software may require additional procedures and/or
   setting changes).
3. Read the Library's Acceptable Use Policy for Wireless Access.
4. Click on **Register** in the left column.
5. Enter the required information.
6. Check your e-mail using your laptop within 60 minutes of registering to get the
   LESSnetworks message and complete the registration process.
7. Start surfing.

Network SSID: Austin Public Library

Printing is not available using the Library's wireless connection. Access a printer by using
the Library's public Internet workstations.

The box is a good eye-catcher.

Need Help?

Be sure your 802.11b wireless card is plugged in tightly to the correct slot. Look for
the light on the card that indicates a signal lock.

Look for the SSID "Austin Public Library" in your wireless card's utility program or in
Windows, go to Network Properties and try "Make New Connection."

Find other wireless access sites in Austin at: *Alternate options.*
http://www.locfinder.com/unitedstates/tx/austin/
http://www.austinwireless.net/ (follow *Nodes Around Austin* link)

Thanks to **WiFi-Texas** and **Schlotzsky's Deli**   *Don't forget to give
for sponsoring the Library's wireless access points.*   *credit where it's due!*

Austin City Connection - The Official Web site of the City of Austin
Contact Us: Send Email or 512-974-7400.
Legal Notices | Privacy Statement
© 1995 City of Austin, Texas. All Rights Reserved.
P.O. Box 1088, Austin, TX 78767 (512) 974-2000

Figure C–11   Sample FAQ. Boulder (CO) Public Library.

 The Standard for Wireless Fidelity.

Wireless Internet Access
Available at the Boulder Public Library

The Boulder Public Library now has wireless Internet access available to patrons at **all** its <u>branches</u>. The following FAQ (Frequently Asked Questions) should assist you in utilizing this free service.

FAQs:
Frequently Asked Questions

Excerpt from:
www.boulder.lib.co.us/general/wireless_faqs.html
See web site for full document.

1. <u>What is wireless?</u>
2. <u>Why wireless?</u>
3. <u>How do I use wireless at the library?</u>
4. <u>Will I need any special settings or passwords to connect?</u>
5. <u>When I access your remote databases, will I have to provide a Library Card?</u>
6. <u>Since I'm using my own equipment, do the general rules about computer use still apply to me?</u>
7. <u>Can the library help me configure my computer?</u>
8. <u>What about virus protection and security?</u>
9. <u>What hours is wireless access available?</u>
10. <u>Is wireless available at the branches?</u>

<u>General Information</u>:
 - <u>Wireless card settings</u>
 - <u>Internet Explorer tips</u>
 - <u>Miscellaneous</u>

3. How do I use wireless at the library?

- The Boulder Public Library System uses the WiFi standard (also known as IEEE 802.11b and 802.11g). These standards provide up to 54 megabits/second connection speed. Speed will vary by location and number of users.

- You will need to bring your own laptop computer to the library and it will need to have built-in WiFi or you will need to install a WiFi network card. The Library does not provide wireless cards.

- Most WiFi equipment will be compatible. However, the library can make no guarantees as to compatibility of your equipment with the library's network.

- Printers are not available to wireless users in the library at this time.

It's good to let your patrons know what differences and similarities there are between using their laptop and using your wired library stations.

5. When I access your remote databases, will I have to provide a Library Card?

- No – for purposes of accessing our databases, you will be treated as if you are on a Patron workstation, and will not have to provide a Library Card.

e-mail: <u>ask@boulder.lib.co.us</u>
Main Branch: 1000 Canyon Blvd., Boulder, Colorado 80302 Ph: (303) 441-3100
(Main entrance and extensive FREE parking located at 11th & Arapahoe.)

 City of Boulder WEB SITE   **Boulder Public Library** WEB SITE

Copyright 2000-2006, Boulder Public Library

Figure C–12 Sample FAQ Chicago (IL) Public Library.

Chicago Public Library   (Screenshot edited for size.)

Home | Search our Catalog | Find It! | View your library card record | Index

WiFi FAQs

Excerpt from: www.chipublib.org/003cpl/computer/wifi/wififaqs.html

What do all the acronyms mean?
AP – Access Point
ISP – Internet Service Provider
SSID – Service Set IDentifier
VPN – Virtual Private Networking
WEP – Wired Equivalent Privacy
WI-FI – Wireless Fidelity
WISP – Wireless Internet Service Provider
WLAN – Wireless Local Area Network

This is a very thoughtful addition.

I don't have a laptop computer. How can I use the network?
Unfortunately, the library does not have laptop computers for loan. You may access the network from computer stations located in other sections of the library.

Does the wireless network pose a health hazard?
No, the wireless network does not pose any health risk. It uses radio signals within the spectrum of safety. While there will always be controversy over the safety of exposure to radio signals, it is something we are exposed to whether we have a wireless network or not.

As you can see, they choose to answer some non-standard, but not unexpected, patron questions.

How long does my connection last while I'm in one of your libraries?
We have not implemented any sort of time limit presently; however, we reserve the right to do so at a later date.

Why can't I use my copy of Outlook/Outlook Express/Eudora/Pegasus Email/AOL or other e-mail clients to send e-mail from my laptop while I'm connected to the Library's WiFi network?
Sending e-mails using a client such as Outlook requires that we open up certain ports on our network. We have decided not to do this because people may try to send "spam" from our library, and unfortunately, it'll look like it was coming from us. Please check with your ISP to see what their web-mail site is and use it to send and receive e-mail while you're on our network.

I have problems connecting with Internet Explorer - IE
In some cases, the proxy server setting is present in your browser. On a public network like the Library network, it's important that you turn off proxy servers. The wireless network cannot allow unauthenticated connections to external proxy servers for security reasons. They give clear instructions on how to check this.

Can a cell phone interrupt my connection?
A cell phone probably won't interrupt your connection, however there are cordless phones and microwave ovens that operate within the frequency range of the CPL Wireless (2.4 GHz and up) that can cause interference with the connection.

I think I got a virus from your Hotspot.
Hotspots do not produce viruses. They come from the Internet, often as attachments to e-mail. It is strongly recommend that all users have virus protection and personal firewall installed on their Laptops.

◄ Back to the Chicago Public Library Home Page

Read, Learn, Discover! 24 HOURS A DAY

Figure C–13 Sample FAQ. Penn State Harrell Library.

Penn State
The George T. Harrell Library

Excerpt from:
www.hmc.psu.edu/library/GenInfo/wirelesspolicy.htm
See site for full document.

Wireless Computing

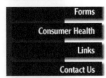

Forms

Consumer Health

Links

Contact Us

The Library Circulation Desk maintains an inventory of wireless network cards for check-out. Alternatively, individuals can use a personal wireless network card that adheres to common industry technical specifications. The privacy of personal and institutional information transmitted wirelessly is safeguarded through required network traffic encryption. Additional information and supported technical specifications are outlined below in section *Wireless Card Technical Specifications.*

(Screenshot edited for size.)

Time Limit for Borrowed Cards

Use of the Cisco Wireless card checked out from the Circulation Desk is limited to 24 hours. Users must sign the Wireless Computing Log Sheet. Users must return Wireless Cards to the Circulation Desk.

Procedure to Obtain A Cisco Wireless Card ⟶ **They lend out wireless cards that can be inserted in a laptop.**

1. Cisco Wireless Cards and an accompanying CD-ROM with installation software are available at the Harrell Library's Circulation Desk.
2. Users must provide a valid HMC or PSU photo ID. NO EXCEPTIONS.
3. Users must sign the log sheet. By signing, users agree to abide by The George T. Harrell Computing Policy.
4. Users will be issued a Cisco Wireless Card and CD with instructions for logging onto and logging off the computer network. The Installation CD needs to be downloaded just one time.
5. The Wireless card and CD must be returned to the circulation desk at the conclusion of the session or after 24 hours which ever occurs first.
6. Users should understand that they are responsible for what is accessed from their computer during the time that the Wireless Card is activated (see HMC Guidelines, Appendix A, and University Policy, Appendix C).
7. Users should understand that they are responsible for the Wireless Card and the installation CD. Lost or stolen cards and CDs will be the responsibility of the user. User will incur replacement costs.

Users may call the Help Desk X6281 for assistance or questions on wireless.

Wireless Printing ⟶ **It is unusual to see libraries offering wireless printing. Clearly, they've found a means to recoup printing costs, even from wireless.**

In order to print to the Pay to Print System, a short script must be executed from a CD onto your laptop computer. The Pay to Print CD may be checked out from the Circulation Desk. The Installation CD needs to be downloaded just one time. Users are responsible for lost or stolen CDs.

Wireless Card Technical Standards

For Windows Operating System 2000
use CISCO AIR-PCM352 PC Card w/Integrated Antenna.

For Windows Operating System XP PRO. XP PRO Home edition is NOT compatible.
Use any wireless card with IEEE 80211.b Network Standard with the following protocols.

- Authentication: 802.1x support including Cisco Leap, and EAP-SIM to yield mutual authentication and dynamic pre-user, per session WEP keys.
- Encryption: Support for static and dynamic IEEE 802.11 WEP keys with 128 bits.

Back

For suggestions or comments about the George T. Harrell website, please email library webmaster.

College of Medicine | Medical Center | Children's Hospital

Privacy and Legal Notices

Penn State Milton S. Hershey Medical Center @2004
This page was last updated on August 26, 2005
Contact Us

SOURCE D: WIRELESS EQUIPMENT MANUFACTURERS

3com

www.3com.com/en_US/jump_page/abg_wireless.html
350 Campus Drive
Marlborough, MA 01752–3064
phone: 508–323–5000 or
1–800-NET–3Com (1–800–638–3266)
fax: 508–323–1111

Alvarion (formerly BreezeCOM)

www.alvarion.com/products/
2195 Faraday Avenue, Suite A
Carlsbad, CA 92008
phone: 760–431–9880
fax: 760–431–8177

Apple Corp.

www.apple.com/airport/
1 Infinite Loop
Cupertino, CA 95014
phone: 408–996–1010

ASUSTeK

usa.asus.com (U.S. website)
phone: 888–678–3688

Belkin Corporation

www.belkin.com
501 West Walnut Street
Compton, CA 90220
phone: 1–800–223–5546

Buffalo Technology, Inc.

www.buffalotech.com/
4030 West Braker Lane, Suite 120
Austin, TX 78759–5319
phone: 512–349–1300 or 1–800–456–9799

Cisco Wireless

www.cisco.com/en/US/products/hw/wireless/
Cisco Systems, Inc.
170 West Tasman Dr.
San Jose, CA 95134
phone: 408–526–4000 or 1–800–553–6387

C-SPEC Corporation

www.c-spec.com
20 Marco Lane
Dayton, OH 45458
phone: 1–800–462–7732 (Don Eshelman) or 937–439–2882
fax: 937–439–2358

Cylink Corporation

www.cylink.com/
P.O. Box 3759
Sunnyvale, CA 94088–3759
Street Address:
910 Hermosa Court
Sunnyvale, California 94086
phone: 1–800–533–3958

D-Link

www.dlink.com
17595 Mt. Herrmann
Fountain Valley, CA 92708
phone: 1–800–326–1688

Enterasys Wireless Solutions

www.enterasys.com/roamabout/
50 Minuteman Rd.
Andover, MA 01810
phone: 978–684–1000 or (877) 801–7082

Glenayre Electronics, Inc.

www.glenayre.com/
5935 Carnegie Boulevard
Charlotte, NC 28209
phone: 704–553–0038

HP Wireless Solutions

www.hp.com/united-states/wireless/
phone: 800–888–5858

Linksys (now a division of Cisco)

www.linksys.com
121 Theory Drive
Irvine, CA 92617
phone: 949–261–1288 or 1–800–546–5797

Netgear Inc.

www.netgear.com
4500 Great America Parkway
Santa Clara, California 95054
phone: 408–907–8000
fax: (408) 907–8097

Proxim, Inc.

www.proxim.com
295 N. Bernardo Avenue
Mountain View, CA 94043
phone: 800–229–1630 or 650–960–1630
fax: 650–960–1984

RadioLAN

www.radiolan.com
4848 San Felipe Road
Box 150–320
San Jose, CA 95135
phone: 1–800–201–2883

Raylink Inc.

www.raylink.com
428 Cloverleaf Drive
Baldwin Park, CA 91706
phone: 626–336–1133 or 1–800–457–6811
fax: 626–336–0517

Solectek Corporation

www.solectek.com
6370 Nancy Ridge Drive
Suite 109
San Diego, CA 92121–3212
phone: 619–450–1220 or
1–800–437–1518 x3035 (Central US & Canada Sales)

U.S. Robotics

www.usrobotics.com
935 National Parkway
Schaumburg, IL 60173
phone: 847–874–2000 or 877–710–0884

Wave Wireless Networking

www.wavewireless.com
1748 Independence Blvd.
Bldg. C–5
Sarasota, FL 34234
phone: 1–800–721–9283 or 941–358–9283
fax: 941–355–0219

Wi-LAN, Inc.

www.wi-lan.com
801 Manning Road NE
Suite 300
Calgary, Alberta
T2E 8J5
CANADA
phone: 800–258–6876 or 403–273–9133
fax: 403–273–5100

SOURCE E: HIGHLIGHTS OF THE MARICOPA COUNTY (AZ) LIBRARY DISTRICT SITE SURVEY

A SITE SURVEY FINAL REPORT

As mentioned in Chapter 3, one major component of WLAN implementation is a comprehensive wireless site survey of your library. You may choose to undertake this yourself, though we strongly recommend that you have it done professionally, either by your organization's IT staff if you have such support, or by an outside vendor. To facilitate this very important step, we wanted to give you an example of a site assessment you might receive back from a wireless vendor. As with the User Experiences we gathered for this book, we approached a fellow librarian for an example of such a report. Vicki Terbovich, Chief Technology Officer of the Maricopa County Library District (MCLD) in Arizona, very kindly offered to share the assessment done for MCLD in December of 2005. The Phoenix-based company that did their site survey graciously agreed to have their report used as an exemplar.

> The company has given me their consent for you to use this as an example of the typical types of information that needs to be gathered in a good wireless assessment.... We want you to encourage libraries if they can afford to do so, to use wireless site assessments so that they can understand how to get the most optimal coverage .... Also, some of this information may be ... [useful to] libraries in conducting their own wireless site assessments. (Conversation with Vicki Terbovich, February 2, 2006)

Excerpts from MCLD's assessment are below, with commentary. We give our very grateful thanks to Vicki, her colleagues at MCLD IT Services, and her site surveyor for their generosity. We hope you will find it useful.

SITE ASSESSMENT CHECKLIST

Before proceeding to the Maricopa County document, let's review what a site survey final report, or site assessment, should contain. This list is also included in Chapter 3, but we felt it would be useful for reference when looking at MCLD's document. Your final site survey report, whether produced by you, your IT department, or an outside vendor should cover the following items:

- If hiring an outside vendor, **information on their company,** including insurance (bond) and experience of the main players in implementing wireless networking.
- A **timeline** for implementation.
- Any **permits required or regulations** that affect the site plan—this is rarely an issue, but if you share space with another company or are in a clinic or hospital building there may be some limitations and regulations to consider.
- What **standard** the network equipment will use (802.11b/g?).
- Any **rogue wireless networks** they detected that might interfere with your network, and how they plan to deal with them.
- **Security** level/type to be used (WEP, WPA, etc. – see Chapter 5).
- A **list of equipment needed,** including access points, routers, hubs, switches, repeaters or wireless bridges. Also if any APs will require high-gain or directional **antennae** (purchased separately) rather than standard omnidirectional.
- A **list of software or firmware** you may need to purchase in addition to what will come with equipment. This would be necessary largely if you plan to have additional layers of security on your network or if they are recommending a package of wireless network management software (see Chapter 4).
- **Brand recommendations** and **price quotes** for the above.
- If they offer such a thing, get an **idea of the lifespan of the above equipment.** Will you need to replace it in 2 years, 3 years? Will it need replacing due to regular wear, or because the vendor sees a new technology coming down the pike that will make it obsolete.
- A **map or maps,** including the following information:
  - o **Locations (tentative) of Access Points** (APs). Will they be in the ceiling, on high shelves, in the open or behind panels?

o **RF lines of sight** – how will the APs communicate with each other through the building, and will you need repeaters in some areas to boost the signal from one AP to the next?

o Your **channel configuration** – which APs will use which of the three primary channels: 1, 6, and 11. A general map of the channel overlap and placement for each building is useful.

o Based on all of the above, what the **expected signal coverage area** will be, with probable *strengths* (e.g., strong signal in your study rooms, but weaker signal out near the windows) and *expected interference* (that pesky elevator shaft). This can be simply shading on the map, to give you a rough idea of your best "hot spots."

o **Any additional electrical or Ethernet cabling** to be run. Remember that you might decide to use PoE (Power over Ethernet) adaptors instead of running both electrical and Ethernet to a location. Have any vendor cost out both options for you.

THE MARICOPA COUNTY LIBRARY DISTRICT SITE SURVEY DOCUMENT

The MCLD wireless site survey looked primarily at five (5) of the district's branches, which were to have WLANs installed. This survey was part of a larger district-wide project to upgrade network and telephony systems, but this report concentrates on WLAN implementation. The entire report is over 50 pages long, so we will simply excerpt points of note, with the primary purpose of highlighting those things that should be included in a good final site assessment.

Figure E–1 Maricopa County Library District Site Survey Document.

WIRELESS LOCAL AREA NETWORK SITE SURVEY

TECHNICAL FINDING AND INFORMATION GUIDE

Version 1, Release 1.1

26 December 2005

The report begins with a clear and well-written overview of wireless in general, including the basics of RF, signal transmission and interference, antennae options and 802.11 standards. We will not reproduce that here, as much of that information is included in the preceding chapters of this book. The inclusion of such a detailed overview is unusual, though noteworthy. Much of the information would be redundant to you the librarian, especially after reading this book; nonetheless if you have to give copies to a board or committee, it's helpful to have this background information clearly laid out. For most site assessments a simple overview of concepts, with definitions of terms and explanations of the equipment to be discussed, would be sufficient and expected. No final report should intend to baffle you with technical jargon, but should instead give clear explanations, definitions and context. If something is unclear, make the surveyor explain and/or rewrite the report.

To begin, the report should tell you **how the survey was conducted.** Did they walk each building or space and take readings of signals? What software and/or hardware did they use? The MCLD report begins the results section with the following:

The following are the compiled results of these [five branch] surveys, and include information in regards to existing RF traffic (if any), recommended number of wireless access points to provide the desired coverage, and the suggested locations of each access point based on optimal client coverage and within the limitations of supporting wired connectivity for each AP device. The site surveys were performed using three standard Cisco 1200 series dual-band access points with 2.2dbi dipole antenna, utilizing the 802.11b/g frequency spectrums to determine RF propagation at each location. Existing RF radiation present at each location was identified using the NetStumbler application version 0.4.0 for W32. The Cisco Aironet Site Survey Tool was used to determine RSSI and AP placement alternatives for each respective location's proposed number of access points.

NetStumbler is a free software tool, widely used, which can be loaded on a laptop and used to find rogue signal or determine signal strength from installed access points.

They go on to describe the survey of the first branch, the George L. Campbell Branch Regional Library, a large building of 130,000 sq. ft. on one level which includes library administrative offices in addition to the library proper. First they checked for **"rogue" signals** that might interfere with transmissions from the access points; these might come from cordless phones, microwaves, other WLANs and so forth.

The existing RF environment when the site survey was performed was found to be moderate, with spurious RF activity detected across the 802.11b/g frequency spectrum on channels 1, 6, and 11. This RF activity will need to be considered during implementation, but should not provide a significant obstacle to deployment.

One aspect that the surveyors had to note throughout their surveys was that Maricopa plans to add not only wireless data networks but also wireless telephony capabilities. With this in mind, they discuss **signal strength and probable throughput:**

Survey access points' initial placement was determined using best practices guidelines for adequate coverage and overlap, and was verified by walking around with monitoring software and noting signal levels at the required coverage locations. The intent was to verify the maximum distances that will maintain adequate signal levels, which for the purposes of the site survey were defined as having a Signal Quality RSSI of –68dbm or lower to be sufficient signal strength to enable operation at the planned data rate of at least 5.5 Mbps to support the requirements of the Cisco 7920 IP Phone devices that are desired as part of the proposed implementation. If the predetermined location of the access point(s) did not meet the desired coverage parameters, repositioning or including additional access points was implemented, and repeat testing was performed.

When the surveyors know in detail *before* they begin their research what your expectations are—e.g., both staff and public access, data or data and telephony, limitations you want to put on the network for bandwidth or usage—their final report can be much more complete and responsive.

The Campbell Branch section of the MCLD report determined that fourteen (14) **access points** would be required to extend signal to all the expected coverage areas, which included most of the 130,000 sq. ft. building. Part of each branch assessment included a standard **coverage map** with each AP noted and its coverage area shaded (see Figure E–4 as one example). In addition—and this is a *great* idea—they took **photos of the interior of the building** and noted all fourteen AP locations on the photos. (see Figures E–2 and E–3) This is immensely useful, because the librarian in charge can instantly see where each device will be mounted, rather than guessing solely from a two-dimensional map whether an AP will be on a wall, in the ceiling or on a high shelf. They also broke the building up into quadrants (A through D), so that as they explained each AP's location, they could also explain **what coverage each AP would provide** to what areas. For each access point placement, they also noted **whether there was both power and network access available**—basically, electrical and cable present or able to be extended to that point. The following are two out of the fourteen placements. Note that "IMO" stands for Information Module Outlet. This and other acronyms and jargon are explained in a **glossary** at the end of the report, another useful addition.

1) Access Point #1: Ceiling/Wall Southeast Corner of Area A lobby – An AP located here will provide WLAN coverage throughout this area west into the self check area and east into the atrium proper. AP allied support requirements (power/network IMO) exist within IEEE standards for this area. (Figure E–2)

Figure E–2   Access Point #1 Placement.

2) Access Point #2: Northeast Study Room Ceiling – This AP will provide coverage for the northern sections of Area A, and radiate south into the atrium, providing supplemental coverage into Area B southern portion of the Stacks room. Allied requirements (power/network IMO) exist within IEEE standards of this mounting location. (Figure E–3)

Figure E–3   Access Point #2 Placement.

In some cases, they found that the power or network connections were not in optimal locations for their desired placement of APs:

> 4) Access Point #4: North Wall, General Stacks Area – An AP located here will provide WLAN coverage North to South across the center of this area, and West to East along the North half of Area B. This AP will provide the main source of RF energy for the Northwest/Northeast portion of this area, providing supplemental coverage to the southwest border of this area as well. The location of this access point may have to be adjusted in order to allow connection to the required allied support infrastructure, with allied support facilities close to this mounting location.

After all fourteen placements were noted, the coverage map was included.

Figure E–4   Campbell Branch Proposed AP Locations and RF Coverage.

Campbell Branch

Similar placement information and coverage maps were created for each of the branches covered. To give you an idea of what it takes to provide coverage, all of the branches listed were single-level structures, with varying square footage. They required different numbers and placements of access points to achieve the desired coverage.

Figure E–5   Branch Square Footage and Access Points Required.

Branch	Square Footage	Access Points Required
George L. Campbell	130,000 sq. ft.	fourteen (14)
Fountain Hills	15,000 sq. ft.	four (4)
North Valley Regional (adjacent to local school)	24,000 sq. ft.	five (5)
Southeast Regional (Gilbert)	66,000 sq. ft.	seven (7)
Northwest Regional (Surprise)	15,000 sq. ft.	five (5)

To give you an idea of the relative coverage placement in one of the smaller branches, Fountain Hills coverage map (approximately 15,000 sq. ft.) looks like this:

Figure E–6   Fountain Hills Branch Proposed AP Locations and RF Coverage.

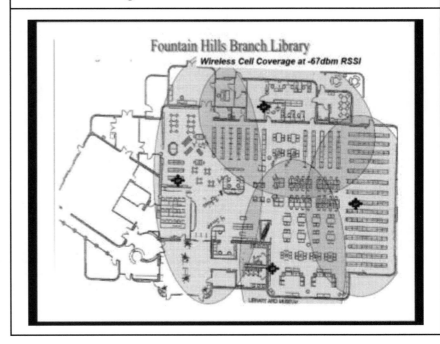

They end the main part of their report with their equipment recommendation for access points, which was for a Cisco product:

> With high-performing dual IEEE 802.11a and 802.11g radios, the Cisco Aironet 1130AG Series provides a combined capacity of up to 108 Mbps to meet the needs of growing WLANs. Hardware-assisted Advanced Encryption Standard (AES) or temporal key integrity protocol (TKIP) encryption provides uncompromised support for interoperable IEEE 802.11i, Wi-Fi Protected Access 2 (WPA2) or WPA security. For the purposes of the MCLD deployment, these APs will use the Lightweight Access Point Protocol (LWAPP). The Cisco Aironet 1130AG Series uses radio and network management features for simplified deployment, along with built-in omnidirec-

tional antennas that provide robust and predictable WLAN coverage for offices and similar RF environments. In addition, when running Cisco IOS Software the Cisco Aironet 1130AG Series supports both access point and workgroup bridge functionality. The competitively priced Cisco Aironet 1130AG Series is ready to install and easy to manage, reducing the cost of deployment and ongoing maintenance.

Again, Maricopa County Library District's needs include more than a basic WLAN setup. They plan to use a WLAN management product from Polaris to centrally manage their wireless connectivity, both for staff and the public. This is not outlined here, though it is understood that it will be a layer on top of this setup. A good vendor will adjust to your requirements and do their best to create the layout you require.

The price list produced for this report was very specific to MCLD's individual needs, including both data and telephony solutions, so we've not reproduced it here. See Chapter 4 for more on pricing. Remember to ask at the outset for options in the final price list—different equipment manufacturers if available, options for different antennae if they would make a coverage difference, etc. A vendor may have a preference for a single manufacturer, which is fine, but they need to be able to explain to you why what they've suggested is the best fit for your organization. Don't be afraid to question the final report and ask for further explanations.

Looking back at our site survey checklist, we see that this report covers all or more of our required elements. They discussed how the survey was conducted, structural issues if any, placement of access points for best signal strength and coverage, channel distribution, and any power or network connection issues those placements might create. For each branch, a test for potentially interfering signals was conducted. They provided the preferred visuals of AP coverage maps and, in a nice addition, photos of the actual placements in the branches themselves. Equipment and security issues were discussed, both in general in the wireless overview at the beginning, and again when recommending a specific model of access point.

A good site survey report or assessment should include these points, but also respond to your organization's particular needs and circumstances. With such a document in hand, you should be able to move quickly and smoothly toward implementing wireless in your library.

INDEX

Page numbers that include definitions are highlighted in bold.

ABOUT THE AUTHORS

Louise E. Alcorn has been the Reference Technology Librarian at the West Des Moines Public Library in Iowa since 1996. She also does library technical editing as a consultant and teaches a professional development class on the reference interview for the Central Iowa Library Service Area. She has worked at the University of Michigan Harlan Hatcher Graduate Library (Special Collections) and at the Kresge Business Administration Library of the University of Michigan Business School.

Louise earned her MILS at the University of Michigan's School of Information in 1995. She also has a B.A. in American Studies from Grinnell College (1992). She was a student research fellow at the Newberry Library in Chicago through the ACM/GLCA Humanities at the Newberry Program in the fall of 1990, assisting in the Atlas of Historical County Boundaries Project. While at the University of Michigan, she was president of the Information and Library Studies Student Association (1995) and participated in the original Internet Public Library Project [www.ipl.org]. Louise is the co-author, with Michael Sauers, of the *Neal-Schuman Directory of Management Software for Public Access Computers* (2003). She has also been published in Web Junction's Technology Resources [www.webjunction.org] and in *Info Career Trends* at LISjobs.com.

Maryellen Mott Allen currently serves as the Coordinator of Instructional Services for the University of South Florida Tampa Library. She has been a librarian for 8 years and has worked in a wide variety of library settings including special, public, and academic. Ms. Allen enjoys conducting research on emerging technologies for information delivery as well as usability and end-user searching behaviors.

135 x 292